SCIENCE
AT THE
Nanoscale

AN INTRODUCTORY TEXTBOOK

SCIENCE
AT THE
Nanoscale

AN INTRODUCTORY TEXTBOOK

Chin Wee Shong • Sow Chorng Haur • Andrew TS Wee

National University of Singapore

PAN STANFORD PUBLISHING

Published by

Pan Stanford Publishing Pte. Ltd.
Penthouse Level, Suntec Tower 3
8 Temasek Boulevard
Singapore 038988

Email: editorial@panstanford.com
Web: www.panstanford.com

British Library Cataloguing-in-Publication Data
A catalogue record for this book is available from the British Library.

ISBN 978-981-4241-03-8 (Hardcover)
ISBN 978-981-4241-17-5 (eBook)

Printed in Singapore.

Acknowledgements

We gratefully acknowledge our students in SP2251 and UPC2206 for their inputs and feedback during the teaching of these courses, the materials of which eventually led to the development of this book. We also thank our graduate students from the Nano-materials Synthesis Laboratory, Colloid Laboratory and Surface Science Laboratory at NUS for the use of their experimental results in the book. Finally, we are grateful to the many laboratory officers, research fellows and colleagues from the NUS Physics and Chemistry departments, and the NUS Nanoscience & Nano-technology Initiative, as well as our families, for their support and understanding. It took two years to put this book together, amidst our heavy commitments.

This page intentionally left blank

Preface

Nanotechnology is one of the most important growth areas in the 21st century. Nanoscience, the science underpinning nanotechnology, is a multidisciplinary subject covering atomic, molecular and solid state physics, as well as much of chemistry. Nanostructures are known to exhibit novel and improved material properties. Fundamentally, these arise because the physical as well as chemical properties are very different when dimensions are reduced to the nanometer range. This book thus aims to introduce the various basic principles and knowledge needed for students to understand science at the nanoscale.

Many ideas proposed in nanotechnology are frontier and futuristic, although some have immediate technological applications. The fundamental scientific principles of all nanotechnology applications, however, are grounded in physics and chemistry.

Nanoscience and nanotechnology degree programmes are being increasingly offered by more and more universities around the world, especially in Australia and Europe.[1] A conventional undergraduate study of a science and engineering discipline normally specialises in the final year(s), but nanotechnology curricula often aim to confront students from their first or second years with the essence and interdisciplinarity of nanoscience and nanotechnology. By introducing the ideas and applications of nanoscience early, students receive a coherent overview of nanoscience to motivate them to learn the necessary basics in the traditional science disciplines of physics, mathematics, chemistry, materials science, biology and medicine. Real interdisciplinarity can be achieved by combining the breadth of nanoscience with the depth in each discipline.

---

[1] Wikipedia (under entry: "Nanotechnology Education", http://en.wikipedia. org/wiki/Nanotechnology_education) states the first programme involving nanotechnology was offered by the University of Toronto, where nanotechnology could be taken as an option within their Engineering Science programme. Interestingly, Wikipedia indicates that to date, Australia leads the world with nine universities offering bachelors degree programmes, followed closely by Europe with about seven

There are some recurring themes in nanoscience and nanotechnology education:

- The *basic foundational disciplines* are crucially important in any nanotechnology program;
- The *nanoscale* has always been important in chemistry, physics, biology and engineering, but with the advent of new tools and technologies, the nanoscale is now openly visible, comprehensible and manipulable;
- It is critical for students to recognize *connections* between the different scientific and engineering disciplines.

The metrics for identifying success in nanotechnology programs are typically the acquiring of multiscale scientific knowledge, and the ability of the programs to keep up-to-date with the latest scientific discoveries.

There are currently numerous specialised nanoscience and nanotechnology-related texts or monographs at the graduate and senior undergraduate levels. This textbook is targeted at the junior undergraduate levels or as a reference text for advanced learners at pre-university and senior high school, and has evolved from the authors' own teaching of the following modules at the National University of Singapore:

- GEK1509 Introduction to the Nanoworld
- UPC2206 Nanoscale Science and Technology
- SP2251 Science at the Nanoscale

The authors' own research expertise cover a diverse range of areas including nanomaterials chemistry and self-assembly (Chin Wee Shong), colloids, nanowires, optical tweezers and atomic force microscopy (Sow Chorng Haur), surface science and scanning tunneling microscopy (Andrew T S Wee). This book therefore aims to be a practical and user friendly textbook that could be adopted in introductory undergraduate courses in nanoscience and nanotechnology, materials science and engineering, physics and chemistry.

Supplementary materials, including solutions to exercises, for this textbook are available at www.panstanford.com/nanotextbook.

Chin Wee Shong, Sow Chorng Haur, Andrew T S Wee
National University of Singapore
2009

About the Authors

Chin Wee Shong is an Associate Professor of Chemistry at the National University of Singapore. Her research interests include the studies of nanostructures and their assemblies, the mechanism of size-, shape- and phase-controlled formation of nanocrystals, solution and templated synthesis of various types of nanomaterials and hybrid materials. She has vast experience in the teaching of undergraduate physical chemistry courses, including topics such as spectroscopy, kinetics, electrochemistry, solid state as well as surface chemistry.

Sow Chorng Haur is an Associate Professor of Physics at the National University of Singapore. His research interests include the studies of nanomaterials systems such as carbon nanotubes and nanostructured metallic oxides, development of nanofabrication techniques such as focused laser beam nanofabrication, studies of colloidal systems, and development of the optical tweezers techniques for contactless manipulation of micro- and nanoscale objects.

Andrew T S Wee is a Professor of Physics at the National University of Singapore. His research interests include surface nanostructure formation, molecular self-assembly on nanotemplates, synchrotron and scanning tunneling microscopy studies of surfaces and interfaces, graphene and related nanomaterials. He is in the editorial board of several journals, including Applied Physics Letters-Journal of *Applied Physics, Surface and Interface Analysis, International Journal of Nanoscience, Surface Review and Letters*, and *Current Nanoscience*. He is also Past President & Fellow of the Institute of Physics Singapore, and a Fellow of the Institute of Physics (UK).

Table of Contents

This page intentionally left blank

Chapter One

Introduction and Historical Perspective

1.1 THE DEVELOPMENT OF NANOSCALE SCIENCE

The prefix *nano* comes from the Greek word for *dwarf*, and hence *nanoscience* (the commonly used term nowadays for nanoscale science) deals with the study of atoms, molecules and nanoscale particles, in a world that is measured in nanometres (billionths of a metre or 10^{-9}, see Section 1.2). The development of nanoscience can be traced to the time of the Greeks and Democritus in 5th century B.C., when people thought that matter could be broken down to an indestructible basic component of matter, which scientists now call *atoms*. Scientists have since discovered the whole periodic table of different atoms (elements, see Section 4.1) along with their many *isotopes*. The 20th century A.D. saw the birth of nuclear and particle physics that brought the discoveries of *sub-atomic particles*, entities that are even smaller than atoms, including *quarks*, *leptons*, etc. But these are well below the nanometre length scale and therefore not included in the history of nanoscale science and technology.

The beginnings and developments of *nanotechnology*, the application of nanoscience, are unclear. The first nanotechnologists may have been medieval glass workers using medieval forges, although the glaziers naturally did not understand why what they did to gold made so many different colours. The process of nanofabrication, specifically in the production of gold nanodots, was used by Victorian and medieval churches which

Science at the Nanoscale: An Introductory Textbook
by Chin Wee Shong, Sow Chorng Haur & Andrew T S Wee
Copyright © 2010 by Pan Stanford Publishing Pte Ltd
www.panstanford.com
978-981-4241-03-8

are famed for their beautiful stained glass windows. The same process is used for various glazes found on ancient, antique glazes. The colour in these antiques depends on their nanoscale characteristics that are quite unlike microscale characteristics.

The modern origins of nanotechnology are commonly attributed to Professor Richard Feynman,[1] who on December 29th 1959 at the annual meeting of the American Physical Society at Caltech, delivered his now classic talk *"There's Plenty of Room at the Bottom"*.[2] He described the possibility of putting a tiny "mechanical surgeon" inside the blood vessel that could locate and do corrective localized surgery. He also highlighted a number of interesting problems that arise due to miniaturisation since "all things do not simply scale down in proportion". Nanoscale materials stick together by molecular van der Waals attractions. Atoms also do not behave like classical objects, for they satisfy the laws of quantum mechanics. He said, "... as we go down and fiddle around with the atoms down there, we are working with different laws, and we can expect to do different things." Feynman said he was inspired by biological phenomena in which "chemical forces are used in repetitious fashion to produce all kinds of weird effects (one of which is the author)". He predicted that the principles of physics should allow the possibility of manoeuvring things atom by atom.

Feynman described such atomic scale fabrication as a *bottom-up* approach, as opposed to the *top-down* approach that is commonly used in manufacturing, for example in silicon integrated circuit (IC) fabrication whereby tiny transistors are built up and connected in complex circuits starting from a bare silicon wafer. Such top-down methods in wafer fabrication involve processes such as thin film deposition, lithography (patterning by light using masks), etching, and so on. Using such methods, we have been able to fabricate a remarkable variety of electronics devices and machinery. However, even though we can fabricate feature sizes below 100 nanometres using this approach, the ultimate sizes at which we can make these devices are severely limited by the

---

[1] Feynman was one of the recipients of the Nobel Prize in Physics in 1965 for his work on quantum electrodynamics. He was also a keen and influential populariser of physics in both his books and lectures.
[2] *Engineering and Science*, Caltech, February 1960.

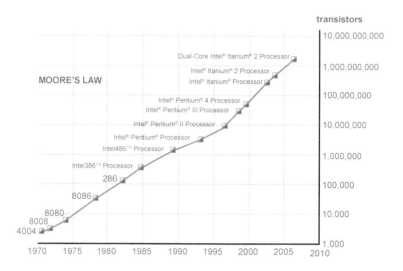

Figure 1.1. Moore's law predicts rapid miniaturization of ICs. [Reprinted with permission from Intel Corporation © Copyright Intel Corporation.]

physical laws governing these techniques, such as the wavelength of light and etch reaction chemistry.

Figure 1.1 shows that the trend in miniaturisation of ICs will ultimately be limited by quantum mechanics, certainly at scales larger than atoms and molecules. Gordon Moore, co-founder of Intel, made the observation in 1965 (now known as "Moore's law") that the number of transistors per square inch on integrated circuits had doubled every year since the integrated circuit was invented. Whilst this trend in IC miniaturisation has more or less been obeyed until now, the current CMOS technology will hit a "wall" soon as quantum and ballistic electron effects become dominant. The most optimistic proponents of ICs believe that major innovations will be required to reach the ultimate operating limit of the silicon transistor: a length for functional features around 10 nm, or about 30 atoms long.

Bottom-up manufacturing, on the other hand, could provide components made of single molecules, which are held together by covalent forces that are far stronger than the forces that hold together macro-scale components. Furthermore, the amount of information that could be stored in devices built from the bottom-up would be enormous.

Figure 1.2. STM image of the Si(111)-7×7 reconstructed surface showing atomic scale resolution of the top-most layer of silicon atoms (from author's lab).

Since Feynman's early visionary ideas on nanotechnology, there was little progress until in 1981 when a new type of microscope, the Scanning Tunneling Microscope (STM), was invented by a group at IBM Zurich Research Laboratory.[3] The STM uses a sharp tip that moves so close to a conductive surface that the electron wavefunctions of the atoms in the tip overlap with the surface atom wavefunctions. When a voltage is applied, electrons "tunnel" through the vacuum gap from the foremost atom of the tip into the surface (or *vice versa*). In 1983, the group published the first STM image of the Si(111)-7×7 reconstructed surface, which nowadays can be routinely imaged as shown in Fig. 1.2.[4] In 1986, Gerd Binnig and Heinrich Rohrer shared the Nobel Prize in Physics "for their design of the scanning tunneling microscope". This invention led to the development of the Atomic Force Microscope (AFM) and a whole range of related Scanning Probe Microscopes (SPM), which are the instruments of choice for nanotechnology researchers today.

---

[3] G. Binnig, H. Rohrer, Ch. Gerber and E. Weibel, *App. Phys. Lett.* **40**, 178–180 (1982).

[4] G. Binnig, H. Rohrer, Ch. Gerber and E. Weibel, *Phys. Rev. Lett.* **50**, 120–123 (1983).

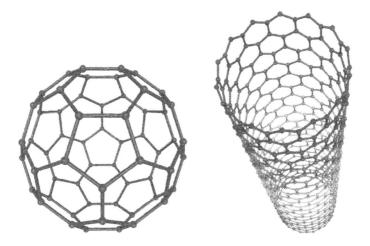

Figure 1.3. Schematic of a C_{60} buckyball (left) and carbon nanotube (right).

At around the same time in 1985, Robert Curl, Harold Kroto and Richard Smalley made the completely unexpected discovery that carbon can also exist in the form of very stable spheres, which they named fullerenes (or *buckyballs*).[5] The carbon balls with chemical formulae C_{60} or C_{70} are formed when graphite is evaporated in an inert atmosphere. A new carbon chemistry has developed from this discovery, and it is now possible to enclose metal atoms in them, and to create new organic compounds. Not long after in 1991, Iijima *et al.* reported Transmission Electron Microscopy (TEM) observations of hollow graphitic tubes or carbon nanotubes, which form another member of the fullerene structural family.[6] The strength and flexibility of carbon nanotubes makes them potentially useful in many nanotechnology applications. Carbon nanotubes are now used as composite fibers in polymers and concrete to improve the mechanical, thermal and electrical properties of the bulk product. They also have potential applications as field emitters, energy storage materials, molecular electronics components, and so on. Some important events in the historical development of nanoscience and nanotechnology are summarised in Table 1.1.

---

[5] H. W. Kroto, J. R. Heath, S. C. O'Brien, R. F. Curl and R. E. Smalley, *Nature* **318**, 162 (1985).
[6] S. Iijima, *Nature* **354**, 56 (1991).

Table 1.1 Some important events in the historical development of nanoscience and nanotechnology.

5th Century B.C.	Democritus and Leucippus, determined that matter was made up of tiny, indivisible particles in constant motion.
1803	English chemist and physicist, John Dalton (1766–1844), developed the first useful atomic theory of matter.
1897	Cambridge physicist J. J. Thomson (1856–1940), proposed that the mysterious cathode rays were streams of particles (later became known as electrons) much smaller than atoms.
1911	Thomson's student, Ernest Rutherford, determined there was a center of the atom, now known as the nucleus, and electrons revolved around the nucleus.
1914	Swedish physicist Niels Bohr, advanced atomic theory further in discovering that electrons traveled around the nucleus in fixed energy levels.
1959	Feynman gives after-dinner talk describing molecular machines building with atomic precision.
1974	Taniguchi uses term "nano-technology" in paper on ion-sputter machining.
1977	Drexler originates molecular nanotechnology concepts at MIT.
1981	Scanning Tunneling Microscopy (STM) invented by Gerd Binnig and Heinrich Rohrer at IBM Zurich.
1985	Buckyball discovered by Robert Curl, Harold Kroto and Richard Smalley.
1986	Atomic Force Microscopy (AFM) invented by Binnig, Quate and Gerber.
1989	IBM logo spelled in individual atoms by Don Eigler at IBM Almaden.
1990	*Nanotechnology*: First nanotechnology journal by Institute of Physics UK.
1991	Carbon nanotube discovered by Iijima at NEC, Japan.
1993	First Feynman Prize in Nanotechnology awarded.
1997	First nanotechnology company founded: Zyvex.
2000	President Clinton announces US National Nanotechnology Initiative.

In summary, the key events in the short history of modern nanotechnology may be described as follows: The vision of nanotechnology was first popularised by Feynman in 1959, when he outlined the prospects for atomic-scale engineering. In 1981, Binnig and Rohrer invented the *scanning tunneling microscopy*, which enabled scientists to "see" and manipulate atoms for the first time. Corresponding advancements in *supramolecular chemistry*, particularly the discovery of the buckminsterfullerenes (or buckyballs) by Curl, Kroto and Smalley gave scientists a whole class of nanoscale building blocks with which to construct a whole range of nanostructures.

1.2 THE NANOSCALE

To start off our discussion on the nanoscale, we first refer to the metric system. The following table gives a summary of the metric system.

Sometimes it is difficult to appreciate the smallness of the nanoscale. It is thus useful to relate the size scale to items that we commonly find in our home. For example, imagine you take a single strand of human hair. The cross section of a human hair is circular in shape (let us assume to be 100 μm in diametre), and imagine you have a very sharp knife. Use the knife to slice the cross section of the human hair into 100 slices with equal width. After which take out one of the 100 slices and use yet another sharp knife to cut the cross section of that single slice into 1000 slices, again with uniform width. If one takes out one of the 1000 slices, the width of the single strip is equal to 1 nanometre!!! The above hypothetical process is illustrated in Fig. 1.4. This is an extremely small size scale and yet there are lots of fascinating phenomena for us to discover.

If one poses a question, how small is a nanometre? Here are some interesting answers:

(1) the diametre of the C_{60} buckyball molecule
(2) half as wide as a DNA molecule
(3) 2 times the diametre of a Rubidium atom
(4) 10 times the diametre of a Hydrogen atom
(5) the de Broglie wavelength of an electron with an energy of 1.5 eV

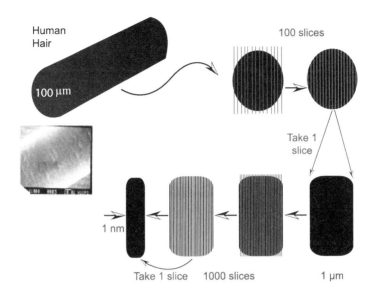

Figure 1.4. Schematic showing systematic cutting down of the cross section of a human hair.

(6) how much your fingernails grow each second
(7) how much the Himalayas rises in every 6.3 seconds
(8) the thickness of a drop of water spread over a square metre

Table 1.2 shows the metric system units, symbols and prefixes relevant to the nanoscale, as well as representative objects at each size scale. Figure 1.5 shows images representing different size scales from one nanometre to one metre.

1.3 EXAMPLES OF INTERESTING NANOSCIENCE APPLICATIONS

(a) Bionanotechnology One of the most exciting areas of applications of nanotechnology must be in the field of biomedical healthcare and disease treatment. The story of tiny "nanobots" acting as *miniaturised doctors* entering our body to repair damaged cells and to kill foreign bacteria alike is certainly not unheard of to many people. While this remains science fiction in many aspects till today, nobody can say for sure that it will never come to pass in the future.

Table 1.2 Metric system units, symbols and prefix.

Name	Abbrev.	Sci. Unit	Unit	Representative Objects with this Size Scale
Yotta-	Ym	10^{24}	1,000,000,000,000,000,000,000,000	The Great Wall, a network of galaxies has a length of 4.7 Ym and a width of 1.8 Ym.
Zetta-	Zm	10^{21}	1,000,000,000,000,000,000,000	Mean diametre of the Andromeda Galaxy is 200,000 lightyears. That is about 2×2 Zm.
Exa-	Em	10^{18}	1,000,000,000,000,000,000	The distance from the Sun to the galactic centre is now estimated at 26,000 light-years ~ 246 Em.
Peta-	Pm	10^{15}	1,000,000,000,000,000	Distance to our nearest star is about 4.3 lightyears ~ 40 Pm.
Tera-	Tm	10^{12}	1,000,000,000,000	The size of our solar system is about 12×10^{12} m.
Giga-	Gm	10^{9}	1,000,000,000	Diametre of Sun is about 1.4×10^{9} m.
Mega-	Mm	10^{6}	1,000,000	The total length of The Great Wall of China is about 5×10^{6} m.
kilo-	km	10^{3}	1,000	Size of Singapore is $=$ about 42 km $\times 23$ km.
hecto-	hm	10^{2}	100	Height of the Great Pyramid of Giza is about 140 m.

Table 1.2 (Continued)

Name	Abbrev.	Sci. Unit	Unit	Representative objects with this size scale
deka-	dam	10^1	10	Length of a type of dinosaur (Apatosorus) $\sim$20 m.
metre	m	10^0	1	Height of a 7-year-old child.
deci-	dm	10^{-1}	1/10	Size of our palm.
centi-	cm	10^{-2}	1/100	Length of a bee.
milli-	mm	10^{-3}	1/1,000	Thickness of ordinary paperclip.
micro-	μm	10^{-6}	1/1,000,000	Size of typical dust particles.
nano-	nm	10^{-9}	1/1,000,000,000	The diametre of a C_{60} molecule is about 1 nm.
pico-	pm	10^{-12}	1/1,000,000,000,000	Atomic radius of a Hydrogen Atom is about 23 pm.
femto-	fm	10^{-15}	1/1,000,000,000,000,000	Size of a typical nucleus of an atom is 10 femtometres.
atto-	am	10^{-18}	1/1,000,000,000,000,000,000	Estimated size of an electron.
zepto	zm	10^{-21}	1/1,000,000,000,000,000,000,000	Really small.
yocto	ym	10^{-24}	1/1,000,000,000,000,000,000,000,000	Really really small.

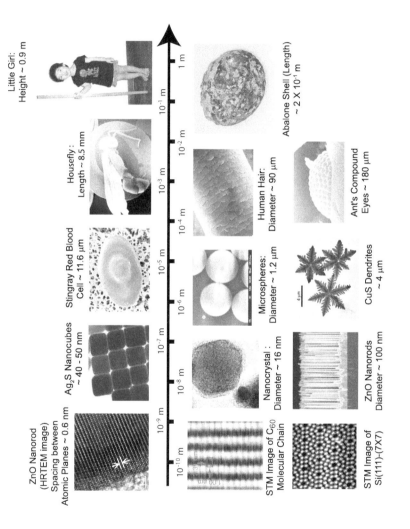

Figure 1.5. Scale of things.

A more promising application of bionanotechnology that has attracted much interest from researchers and industries is the development of nano-drug delivery systems. In our modern busy lifestyle, administration of drugs has progressed from the teaspoon to time-release capsules or implants. Nanotechnology promises delivery mechanisms that can administer drugs at desired rates and at the exact location in the body. This requires the fabrication of precise nanostructures for drug-eluting coatings, membranes, or even implants, For example, researchers at the University of California, San Francisco have demonstrated how they can use nanotubes made from biocompatible metal oxides to hold therapeutic drugs and deliver these agents in a highly-controlled manner.[7] On the other hand the dendrimer, a highly branched polymer, has also been investigated by many as a natural form of nanoparticle carrying myriad sites for drug loading. All these developments not only translate to time-saving and better treatments, they also help avoid side effects caused by large doses taken orally or by injection. There are also the potential benefits of extension of the bioavailability and economic life-span of proprietary drugs. According to the industry consulting firm NanoMarkets, nanotechnology-enabled drug delivery systems are expected to generate over US$1.7 billion in 2009 and over $4.8 billion in 2012.[8]

Another development in nanoscience that has excited many biomedical researchers is the use of quantum dots (abbreviated QDs, see Section 6.1) in bio-imaging. These are tiny crystals that give strong fluorescence signals and, when injected into cells, allow unprecedented details inside the cells to be imaged. A nice 3D imaging example was demonstrated by Cornell researchers (Fig. 1.6) whereby tiny blood vessels beneath a mouse's skin were viewed with CdSe/ZnS QDs circulating through the bloodstream. The images appear so bright and vivid in high-resolution that researchers can see the vessel walls ripple at 640 times per minute.

(b) Spintronics For many years, scientists and engineers have created a host of electrical devices that rely on electrons in the materials. Such devices include the ubiquitous transistor and the

---

[7] C. C. Lee, E. R. Gillies, M. E. Fox, S. J. Guillaudeu, J. M. Fréchet, E. E. Dy and F. C. Szoka, *Proc. Natl. Acad. Sci.*, USA, **103**, 16649–54 (2006).
[8] The NanoMarkets report 2005/03 on *Nano Drug Delivery*: http://www.the-infoshop.com/study/nan24488_nano_drug_delivery.html

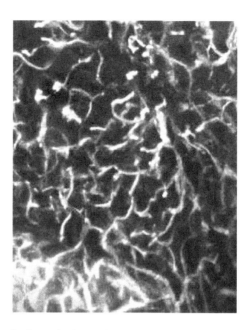

Figure 1.6. The branched capillary structure, feeding adipose tissue in a living mouse, is revealed with multiphoton fluorescence microscopy as nanocrystal quantum dots circulate through the bloodstream. [From Larson D. T. *et al.*, *Science* **300**, 1434–1436 (2003). Reprinted with permission from AAAS.]

powerful microprocessor. These devices exploit the charge carried by the electrons for their normal function, and they communicate with each other through the flow of electric charges. However, there is another important intrinsic property of electrons that has been neglected in these devices — the spin of the electron. Spin is a purely quantum mechanical property. We normally think of the spin of an electron using the analogy of a spinning top. The spin can be clockwise or counterclockwise in direction. In the case of electrons, the spin could be pointing in the "up" direction or in the "down" direction. The spin in the electron is easily influenced by an externally applied magnetic field. Spin electronics, or *spintronics*, refers to electrical devices that utilise the spin properties of the electrons in addition to their electrical charge in creating useful devices. Scientists and engineers hope to control the spin of electrons within a spintronics device to produce useful devices. As the spintronics device can be influenced by the presence of an electric field, magnetic field or light, the device represents a single

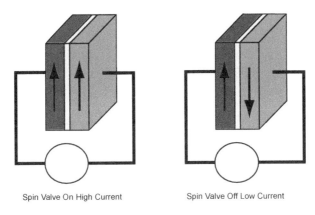

Spin Valve On High Current   Spin Valve Off Low Current

Figure 1.7.   Schematic diagram of a simple spin valve.

device that integrates the multiple functionalities with optoelectronics and magnetoelectronics.

There are a number of spintronics devices that have been realised. The most widely used spintronics device is the Giant Magnetoresistive (GMR) device commonly used in magnetic hard-disk drives. Typically, a simple GMR device consists of two layers of ferromagnetic materials separated by a very thin spacer layer which is nonmagnetic. A simple illustration of such a spin valve device is shown in Fig. 1.7. One of the layers is referred to as the "pinned" layer where its magnetisation direction remains in a fixed direction. The other ferromagnetic layer is known as the "free" layer where its magnetisation direction depends on the externally applied magnetic field. When the two magnetisation vectors of the ferromagnetic layers are oriented in the same direction, an electrical current will flow freely. On the other hand, if the magnetisation vectors are oriented in the opposite direction, there is a high resistance to the flow of electrons due to spin dependent scattering. The magnitude of the change in the resistance at these two different states is called the Giant Magnetoresistance Ratio. Hence this GMR device is highly sensitive to the external magnetic field which is capable of switching the relative magnetic orientation of the ferromagnetic layers. Thus it is widely used as the read head for magnetic hard disk drives.

There are many other spintronics devices that scientists and engineers are working on. These include the spin-based transistor, spin-polarizer, spintronics solar cell, magnetic tunnel junction, and spin-based quantum computer where the spin of a single electron trapped in a quantum dot is used as a *qubit*.

(c) Molecular electronics The emerging field of molecular electronics is now becoming a popular alternative paradigm to current silicon microelectronics. In 1974, Ari Aviram and Mark Ratner, then at New York University, published a paper in *Chemical Physics Letters* proposing that individual molecules might exhibit the behaviour of basic electronic devices.[9] Their hypothesis, formulated long before anyone was able to test it, was so radical that it was not pursued for another 15 years. The story continues in December 1991, when James Tour and Mark Reed discovered they had a common interest at a small gathering of "moletronics" researchers in the Virgin Islands. The meeting was hosted by Ari Aviram, who was then working at IBM's Thomas J. Watson Research Center in New York. They started collaborating, but it was not until 1997 when they successfully used the so-called "break-junction" technique to measure the conductance of a single molecule.[10] In their work, benzene-1,4-dithiol molecules were self-assembled onto two facing gold electrodes of a mechanically controllable break junction to form a stable gold-sulphur-aryl-sulphur-gold system (Fig. 1.8). This allowed the direct observation of charge transport through the molecules for the first time. Their study provided a quantitative measure of the conductance of a junction containing a single molecule, which is a fundamental step towards the realization of the new field of molecular electronics.

Many papers have since followed demonstrating conductance measurements on single molecules and simple single molecule devices. A useful review of the early days of the field has been written by Carroll and Gorman.[11] Nanogaps were formed using electromigration whereby a high electric field causes gold atoms to move along the current direction, eventually causing a nanogap. More recently, a research team at Hewlett-Packard (HP) Laboratories has proposed the crossbar architecture as the most likely path forward for molecular electronics.[12] A crossbar

---

[9] A. Aviram and M. A. Ratner, *Chem. Phys. Lett.* **29**, 277–283 (1974).

[10] M. A. Reed, C. Zhou, C. J. Muller, T. P. Burgin and J. M. Tour, *Science* **278** (1997) 252.

[11] R. L. Carroll and C. B. Gorman, *Angew. Chem. Int. Ed.* **41**, 4378 (2002).

[12] P. J. Kuekes, G. S. Snider and R. S. Williams, *Scientific American*, November 2005, 72.

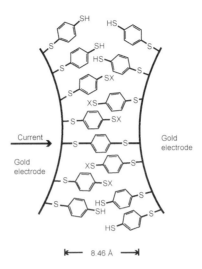

Figure 1.8. Schematic of the gold-sulphur-aryl-sulphur-gold system. [From Reed *et al.*, *Science*, **278**, 252 (1997). Reprinted with permission from AAAS.]

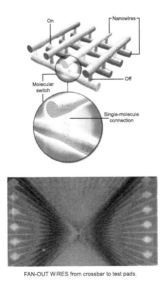

Figure 1.9. Left: Schematic showing how the switch is formed at the junction between two crossing nanowires that are separated by a single monolayer of molecules. Right: Picture of fan-out wires that connect the nanoscale circuits to the microscale. [Reprinted with permission from P. J. Kuekes, G. S. Snider and R. S. Williams, *Scientific American*, November 2005, 72. Copyright © 2005 by Scientific American, Inc. All rights reserved.]

consists of one set of parallel nanowires less than 100 atoms wide that cross over a second set (Fig. 1.9). A molecule or material that can be stimulated electrically to conduct either more electricity or less is sandwiched between the two sets of wires. The resulting interwire junctions form a switch at each intersection between crossing wires that can hold its "on" or "off" status over time. Such switches may be able to scale down to nearly single-atom dimensions, and this approach suggests how far the future miniaturisation of ICs might someday go.

Further Reading

Feynman's 1959 talk at the annual APS meeting in Caltech: "There's Plenty of Room at the Bottom": http://www.zyvex.com/nanotech/feynman.html

Richard Booker, Earl Boysen, Erik Haroz, Earl Boysen, "*Nano technology for Dummies*" (Wiley, 2005).

Eric K. Drexler, "*Engines of Creation: The Coming Era of Nanotechnology*" (Knopf, 1987).

Gerber C. and Lang H. P., "*How the Doors to the Nanoworld were Opened*", *Nature Nanotechnology* **1** (2006) 3.

S. A. Wolf *et al.*, "*Spintronics: A Spin-Based Electronics Vision for the Future*", *Science* **294**, 1488–1495 (2001).

Sankar Das Sarma, "*Spintronics*", *American Scientist* **89**, 516–523 (2001).

Mark A. Ratner, "*Introducing Molecular Electronics*", *Materials Today*, February 2002.

This page intentionally left blank

Chapter Two

Classical Physics at the Nanoscale

It is often said that "physics is different at the nanoscale". This statement cannot be true, since the laws of physics as we know them today are certainly valid at the nanoscale. Perhaps what is often meant is that new phenomena or "new physics" are often observed when we build novel structures, materials and devices at the nanoscale.

Nevertheless, it is true that at the nanoscale, classical physics begins to give way to quantum physics in terms of description of physical phenomena. When we try to describe the properties of electrons, classical physics fails and we have to use the quantum mechanical wave description of matter to explain the physics observed. The length scale of electrons is in any case much smaller than that of atoms and molecules, even though they determine many materials properties such as conductivity, magnetism and so on. This subject of quantum physics will be dealt with in the next chapter.

At the scale of nano- and micro-particles however, we can adequately describe many physical phenomenon with classical physics. We often ask questions such as: Why do dust particles float in the air instead of falling to the ground? Why does a small drop of water not spread but remain round? Why do micron-sized wheels have so little inertia? At this scale, the behaviour of objects is different from what we experience in our daily lives. This is because at the small scale, forces such as friction and surface tension often dominate over forces such as gravity.

Science at the Nanoscale: An Introductory Textbook
by Chin Wee Shong, Sow Chorng Haur & Andrew T S Wee
Copyright © 2010 by Pan Stanford Publishing Pte Ltd
www.panstanford.com
978-981-4241-03-8

In the following sections, we will discuss how some important physical properties such as mechanical frequency, viscosity and motion of nanoscale objects differ from those of macro-sized objects that we normally see.

2.1 MECHANICAL FREQUENCY

A cantilever is a beam anchored at one end and projecting into space. Cantilevers are widely found in construction, notably in cantilever bridges and balconies, as well as in aircraft wings. Civil and aircraft engineers are very concerned about the mechanical frequencies that these cantilever structures are subjected to, since external frequencies around the resonant frequencies of these structures can lead to catastrophic failures and major disasters.

Cantilevered beams are very often found in micro-electro-mechanical systems (MEMS). MEMS cantilevers are commonly fabricated from silicon, silicon nitride or polymers. The fabrication process typically involves undercutting the cantilever structure to release it, often with an anisotropic wet or dry etching technique. In particular, the important technique of atomic force microscopy (AFM) depends on small cantilever transducers. Other applications of micron-scale MEMS cantilevers are in biosensing and radio frequency filters and resonators.

Mechanical resonance frequencies of cantilevers depend on their dimensions; the smaller the cantilever the higher the frequency. To do a simple analysis of the mathematical size

Figure 2.1. (left) Cavenagh bridge, Singapore's oldest suspension (cantilever) bridge; (right) SEM image of an AFM cantilever (from author's lab).

dependence, let us consider the simple case of a mass m attached to a spring with spring constant k. From Hooke's law, when a spring is slightly displaced in the direction x from its equilibrium position, it would undergo simple harmonic motion according to:

$$F = -kx \tag{2.1}$$

From Newton's Second Law:

$$F = m\frac{d^2x}{dt^2} = -kx \tag{2.2}$$

$$\frac{d^2x}{dt^2} = -\frac{k}{m}x = -\omega^2 x \tag{2.3}$$

where $\omega = \left(\frac{k}{m}\right)^{1/2}$, which is the frequency of the sinusoidal equation that is a solution of equation (2.3):

$$x = A\cos(\omega t + \varphi) \tag{2.4}$$

where A is the amplitude of oscillation and ϕ is the phase of the oscillation.

The frequency of oscillation f is the inverse of the period of oscillation T:

$$f = \frac{1}{T} = \frac{\omega}{2\pi} = \frac{1}{2\pi}\left(\frac{k}{m}\right)^{1/2} \tag{2.5}$$

Since the mass and spring are three-dimensional, the mass m will vary as L^3 and the spring constant k as L:

$$\omega \propto {}^1\!/_L \tag{2.6}$$

Hence the frequency is inversely proportional to the length scale for a mechanical oscillator. Frequencies inversely proportional to the length scale are typical of mechanical oscillators such as string instruments like the violin or harp. In such oscillators with two nodes at both ends (i.e. fixed at both ends), the length of the string (oscillator) is related to its lowest order standing wavelength by $L = \lambda/2$. If the oscillator has only one node at one end (i.e. fixed at one end) as in a cantilever, then $L = \lambda/4$. Since $\lambda = vt$, where t is the time for the wave to travel one oscillation and v the wave velocity, the frequency can be written as:

$$\omega = \frac{2\pi}{t} = \pi\left(\frac{v}{L}\right) \text{ (nodes at both ends)} \tag{2.7}$$

$$\omega = \frac{2\pi}{t} = \pi\left(\frac{v}{2L}\right) \text{ (node at one end)} \tag{2.8}$$

From the physics of waves, the wave velocity v can be expressed as:

$$v = \sqrt{\frac{T}{\rho}} \tag{2.9}$$

where T is the tension of the stretched string and ρ is its mass per unit length. For a three dimensional solid material, we can write the wave velocity in (2.8) in terms of the Young's modulus Y of the material (Y = force per unit area per fractional deformation [Pa], or stress/strain), and the material density ρ (kgm^{-3}):

$$v = \sqrt{\frac{Y}{\rho}} \tag{2.10}$$

From Table 2.1, for a silicon cantilever with $Y = 182$ GPa and $\rho = 2300$ kgm^{-3}, we can calculate the speed of sound in silicon to be $v = 8900$ ms^{-1}. From equation (2.8), the resonant frequency of a 1 m long silicon cantilever is $\omega = 14$ kHz. If we reduce the length of the silicon cantilever to 1 cm, its resonant frequency will be about 1400 kHz. A typical silicon AFM cantilever with k between 0.01–100 N/m has a resonant frequency ω of 10–200 kHz. It can be

Table 2.1 Elastic properties of selected engineering materials.

Material	Density (kg/m^3)	Young's Modulus (GPa)
Diamond	1800	1050
Silicon nitride	2200	285
Steel	7860	200
Silicon	2300	182
Aluminum	2710	70
Glass	2190	65
Polystyrene	1050	3

shown (left as an exercise for the reader) that for a thin rod made of a material with atomic spacing a and atomic mass m:

$$v^2 = \frac{Y}{\rho} = \frac{ka^2}{m} \tag{2.11}$$

If the silicon cantilever is further reduced to 1 μm in length, the resonant frequency will be about 14 MHz. Carr *et al.* have measured the resonant frequencies of silicon nanowires and found it to be 400 MHz for a 2 μm long nanowire.[1] If such nanowires had lengths in the nanometre range, they will have resonant frequencies in the GHz range, which would have novel device applications.

The upper limit to oscillation frequencies will be those of molecular bonds. For molecular bond vibrations with bond lengths of about 1-2 Å, the frequencies are of the order of 10^{13} Hz. This is routinely measured by chemists using Infrared (IR) Spectroscopy, a typical spectrum of which is shown in Fig. 2.2. Note that chemists

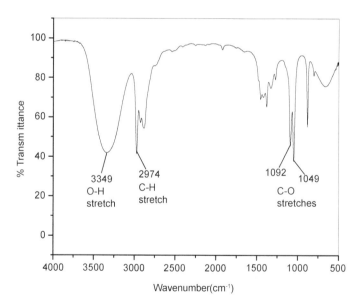

Figure 2.2. FT-IR spectrum of ethanol molecules in liquid state (note: Wavenumber cm^{-1} = Hz/c) (from author's lab).

---

[1] D. W. Carr, *Appl. Phys. Lett.* **75**, 920 (1999).

typically use the unit of wavenumbers (cm^{-1}), and if we convert the wavenumber of the C-H stretch ($2974\,cm^{-1}$) to frequency (Hz), we get about 9×10^{13} Hz. Hence, at the nanoscale, mechanical frequencies are much higher than those of objects at larger scales.

2.2 VISCOSITY

The force F needed to move a sphere of mass m, density ρ, radius R at a velocity v through a viscous medium of viscosity η (Stoke's Law) is given by:

$$F = 6\pi\eta Rv \qquad (2.12)$$

When the sphere reaches terminal velocity v_t, the force on it due to gravity ($F = mg$) is balanced by the retarding force due to the viscosity of the medium:

$$v_t = \frac{mg}{6\pi\eta R} = \frac{\frac{4}{3}\pi R^3 \rho g}{6\pi\eta R} = \frac{2\rho g R^2}{9\eta} \propto R^2 \qquad (2.13)$$

Since the terminal velocity is proportional to the radius squared, it is clear that small particles fall very much more slowly. Note that the above treatment is only valid under conditions of streamline flow, for small particles and low velocities. This condition is met when the Reynolds Number (Re) is less than about 2000, where Re is a non-dimensional quantity that describes the type of flow in a fluid defined by:

$$Re = \frac{2R\rho v}{\eta} = \frac{Inertial \cdot forces\,(\rho v)}{Viscous \cdot forces\,(^{\eta}/_{2R})} \qquad (2.14)$$

As size decreases, the ratio of inertia forces to viscous forces within the fluid decreases and viscosity dominates. Hence, micro/nano-scale objects moving through fluids are dominated by viscous forces, and their motion is characterised by a low Reynolds number. This means that nanoparticles "feel" the viscosity (or 'gooeyness') of the fluid much more than we do!

To give a quantitative example, consider an iron sphere of radius 1 mm and density $7,000\,kgm^{-3}$ (i.e. a small ball bearing) falling through water ($\eta = 0.01$ Pa.s, cf. Table 2.2). It has a terminal velocity calculated from Eq. (2.13) of about 1 ms^{-1}. If the sphere is now 1 μm in radius, its terminal velocity becomes

Table 2.2   Viscosities of some common fluids.

Fluids	Viscosity (Pa.s)
Acetone	0.0032
Air	0.00018
Alcohol (ethyl)	0.012
Blood (whole)	0.04
Blood plasma	0.015
Gasoline	0.006
Glycerine	14.9
Oil (light)	1.1
Oil (heavy)	6.6
Water	0.01

about 1 μms^{-1}, i.e. it falls by a distance equivalent to its size every second. If its radius is further reduced to 1 nm (i.e. an iron nanoparticle), its terminal velocity drops to 1 pms^{-1}, which is negligible relative to its size! Furthermore, at the nanoscale, we expect the effects of individual molecules in the fluid impacting on the nanoparticle (Brownian motion) to become significant, and this will be discussed next.

2.3   BROWNIAN MOTION OF NANOSCALE OBJECTS

In 1827, the English botanist Robert Brown noticed that pollen grains suspended in water jiggled about under the lens of the microscope, following a zig-zag path like the one pictured in Fig. 2.3. It was only in 1905 when Einstein succeeded in stating the mathematical laws governing the movements of particles on the basis of the principles of the kinetic-molecular theory of heat. According to this theory, microscopic bodies suspended in a liquid perform irregular thermal movements called Brownian molecular motion. Brownian motion became more generally accepted because it could now be treated as a practical mathematical model. Its universality is closely related to the universality of the normal (Gaussian) distribution.

The 1D diffusive Brown motion probability distribution as a function of position x and time t, $P(x,t)$, is described by the

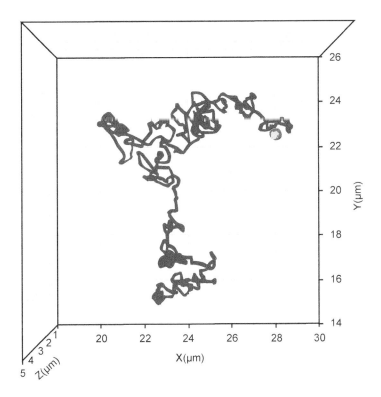

Figure 2.3. A microsphere that is suspended in water will exhibit Brownian motion due to frequent collisions with water molecules. The figure shows the reconstructed trajectory of such a microsphere (diameter = 0.6 μm) after its motion was tracked by optical microscopy over a period of 20 seconds. The zig-zag path is typical of a particle exhibiting Brownian motion (from author's own work).

Gaussian distribution:

$$P(x,t) = (4\pi Dt)^{-3}\, 2\exp\left(\frac{-x^2}{4Dt}\right) \tag{2.15}$$

where D is the diffusivity of a particle radius of R in a fluid of viscosity η at temperature T:

$$D = \frac{kT}{6\pi\eta R} \tag{2.16}$$

From the Gaussian relation in (2.15), we can define the characteristic 2D diffusion length as:

$$x_{rms} = (4Dt)^{1/2} \qquad (2.17)$$

If we go back to the example in the previous section of the iron nanoparticle of 1 nm radius falling through water with terminal velocity of 1 pms^{-1}, the corresponding Brownian diffusion length $x_{rms} = 2D^{1/2}$ is about 9 μm, which is the characteristic distance displaced every second due to Brownian motion. This value is much larger than 1 pm, and hence Brownian diffusive motion is dominant for the 1 nm particle.

If however the iron particle radius was 1 μm, its diffusion length is now 0.3 μm, which is almost comparable to its terminal velocity of 1 μms^{-1}. Hence both diffusive motion and viscosity of the fluid need to be taken into account in describing the particle's motion. In general, Newton's law of motion in such cases in the presence of an external force F_{ext} and taking into account the Brownian diffusive force $F(t)$ and viscosity η of the fluid can be written in 1D as:

$$F_{ext} + F(t) - (6\pi\eta R)\frac{dx}{dt} = \left(\frac{4\pi R^3 \rho}{3}\right)\frac{d^2x}{dt^2} \qquad (2.18)$$

This so-called *Langevin equationLangevin equation* is a *stochastic* differential equation in which two force terms have been added to Newton's second law: One term represents a frictional force due to viscosity, the other a *random* force $F(t)$ associated with the thermal motion of the fluid molecules. Since friction opposes motion, the first additional force is proportional to the particle's velocity (dx/dt) and is oppositely directed. This equation needs to be solved to describe the complete motion of a nanosized-object in a fluid.

2.4 MOTION AT THE NANOSCALE

It has been often hypothesised that in the not-too-distant-future, micron-sized *medical nanorobots* will be able to navigate through our bloodstream to destroy harmful viruses and cancerous cells (see Figure 2.4). This is reminiscent of the 1966 science fiction film *Fantastic Voyage* written by Harry Kleiner, which was

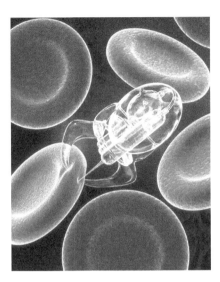

Figure 2.4.   Artist's impression of a nanobot killing a virus.

subsequently written as a novel of the same title by Isaac Asimov, based on the screenplay. *Fantastic Voyage II: Destination Brain* was written by Isaac Asimov as an attempt to develop and present his own story apart from the 1966 screenplay. *Fantastic Voyage: Microcosm* is a third interpretation, written by Kevin J. Anderson, and published in 2001. This version updates the story with modern ideas of nanotechnology, but uses the same theme of miniaturising a crew of scientists, doctors and technicians to investigate a body.

We have seen from the earlier sections that the motion of a nanobot in a fluid would be complex and difficult to control. The viscosity of the fluid is greatly enhanced at the nanoscale making design of the propulsion system a major engineering challenge. Brownian motion would cause a constant random shaking that would also make engineering design difficult. Furthermore, surface forces at the nanoscale become significant, resulting in the nanobot sticking to any surface it comes into contact with.

Nevertheless, we can use these effects to our advantage by getting inspiration from Nature. After all, living organisms such as viruses and bacteria are able to find their way into human cells. If we can design molecules with sticky and non-sticky areas, then the agitation caused by Brownian motion will eventual lead to molecules sticking together in very well-defined ways to form rather complex macromolecular structures. This mode of assembly is known as self-assembly, and will be discussed further in Chapter 7. Such random stochastic processes are the basis of all chemical reactions, and indeed of the biochemistry of life itself.

So far we have been describing physics at the nanoscale using purely classical physics. However, quantum mechanical effects become significant when we consider even smaller entities such as the electron. Indeed changes in energy levels occur when electrons are confined to nano-sized objects, altering the electronic and optical properties of the material. We shall address these issues in the next chapter.

Further Reading

Edward L. Wolf, *Nanophysics and Nanotechnology* (Wiley, Germany, 2004).

Richard A. L. Jones, *Soft Machines — Nanotechnology and Life* (OUP, 2004).

W. R. Browne, B. L. Feringa, *"Making molecular machines work"*, *Nature Nanotechnology* **1**, Oct 2006, 25.

Exercises

1. A steel bridge spanning a river is 100 m long and fixed only at the two ends. Calculate (i) the speed of sound in the bridge; (ii) the resonant frequency of the bridge. Can a class of students oscillate this bridge by jumping on it in a coordinated matter? (iii) If this steel bridge is 1 km long instead, what might happen if a battalion of soldiers march in step across it? (iv) If a micro-model of this steel bridge is made 1μm long, what would be its fundamental frequency, and the next two harmonics?

2. (i) Estimate the terminal velocity of a skydiver falling from a plane. State all assumptions made. (ii) The terminal velocity of a skydiver has actually been measured to be about 200 km/h (or 55 m/s). For a heavy object, the air resistance is proportional to the falling body's velocity squared (i.e. cv^2, where c is a constant). Using this information, determine the value of c and write down the equation of motion for the skydiver of mass 70 kg. (iii) For a bug 100 μm in size, estimate its terminal velocity in air. Assume the bug is just able to float in water.

3. Calculate the Brownian diffusion length for a spherical nanobot of radius 200 nm inside the bloodstream of a living human being. Describe its motion inside the bloodstream, assuming the nanobot has no internal propulsion motor. Note that blood velocity can be as high as 1 m/s in the aorta and <1 mm/s in the capillaries.

4. Show that for a thin rod made of a material with Young's modulus Y, density ρ, spring constant k, atomic spacing a and atomic mass m:

$$\frac{Y}{\rho} = \frac{ka^2}{m} \tag{2.19}$$

Hint: The connection between the macroscopic and nanoscopic quantities can be made by considering a linear chain of N masses m separated by springs with spring constant k and length a.

Chapter Three

Brief Review of Quantum Mechanics

With the advent of nanoscience, scientists are creating nanosystems with ever reducing size scales. In some cases, such as quantum dots with a diameter of a few nanometres, the number of atoms in these systems falls in the range of 100 to 100,000 atoms. As we approach the atomic scale, the quantum nature of the nanosystems becomes dominant. Quantum physics encompasses those laws of physics we use to describe and predict the properties of matter at the length scales of atoms and electrons. An understanding of quantum physics is therefore important for understanding the behaviour of the nanomaterials and nanodevices as their dimensions are reduced towards atomic sizes. The main focus of this chapter is to describe some of the basic concepts of quantum physics.

3.1 BASIC QUANTUM PHYSICS AND QUANTUM CONFINEMENT

The first quarter of the 20th century saw the rapid development of quantum physics. During this period, there was a series of groundbreaking experiments that produced observable phenomena and results that could not be accounted for by classical theories of physics. It was during this period that some of the most fundamental and revolutionary concepts of quantum physics were proposed. A brief summary of the important series of experiments is presented in this section.

Science at the Nanoscale: An Introductory Textbook
by Chin Wee Shong, Sow Chorng Haur & Andrew T S Wee
Copyright © 2010 by Pan Stanford Publishing Pte Ltd
www.panstanford.com
978-981-4241-03-8

3.1.1 Blackbody Radiation

Blackbody radiation refers to electromagnetic (EM) radiation emitted from a small hole in a cavity with walls maintained at a certain temperature. A schematic of the experiment is shown in Fig. 3.1(a). Atoms comprising the wall are continuously emitting electromagnetic radiation as well as absorbing radiation emitted by other atoms of the wall. Thus the cavity is filled with electromagnetic radiation. At equilibrium, the amount of energy emitted is equal to the amount of energy absorbed. And the energy density of the EM field is constant. A small hole in the cavity allows some EM radiation to escape from the cavity and be detected during the experiment. The resultant monochromatic energy density of the blackbody radiation as a function of the wavelength typically obtained during the experiment is shown in Fig. 3.1(b). It can be seen from the plot that for a certain temperature, the energy density shows a pronounced maximum at a certain wavelength known as the Wien's wavelength. The Wien's wavelength decreases as the temperature increases. This explains the change in color of a radiating object as its temperature changes.

According to classical theory, the radiated intensity (I) of the emitted radiation is given by

$$I(\lambda, T) \ \propto \ \frac{1}{\lambda^4} \tag{3.1}$$

where λ is the wavelength of the emitted radiation. Hence classically the radiated intensity of electromagnetic radiation is

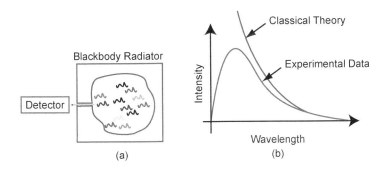

(a) (b)

Figure 3.1. (a) Schematic of the Blackbody radiation experiment and (b) a plot of the energy density of the blackbody radiation at a certain temperature together with the theoretical prediction from classical theory.

expected to increase with decreasing wavelength as shown in Fig. 3.1(b). Obviously, classical theory cannot explain what is observed in the experiment. The motivation to find a mechanism to account for the observation led to the birth of quantum physics.

3.1.2 Max Planck's Theory

In order to provide a good explanation of blackbody radiation, Max Planck proposed the following assumptions:

(a) Atoms of the blackbody radiator behave as harmonic oscillators. The energy of the oscillators adopt the form of discrete values of

$$E = nh\upsilon \tag{3.2}$$

where n is a positive integer and υ is the frequency of the oscillators. Here h is a new fundamental constant of nature known as Planck's constant, $h = 6.63 \times 10^{-34}$ Joule sec.

(b) Each atom can absorb or emit radiation energy packet by going through transition from one state ($E = nh\upsilon$) to an adjacent energy state ($E = [n \pm 1] h\upsilon$). Hence the amount of energy absorbed or emitted by the atom is equal to $h\upsilon$.

The above condition deviates from classical theory and implies that the energy of atomic oscillators is quantised. Such quantisation was subsequently incorporated into other physical quantities and became a fundamental property of many systems in nature. Using this simple but revolutionary assumption, together with the concepts from statistical mechanics, Planck was able to obtain an expression for the energy density in the blackbody radiation that agrees surprisingly well with the experimental observations.

3.1.3 Photoelectric Effect

Photoelectric effect refers to the emission of electrons from a material under the action of light irradiation. The emitted electrons are known as the photoelectrons. The following are a summary of the observations made during the photoelectric experiments.

(a) Emission of the photoelectrons depends on the frequency f of the incident light.

(b) Light with higher frequency gives rise to photoelectrons with higher maximum kinetic energy.

(c) A more intense light source gives rise to larger number of photoelectrons but the maximum kinetic energy of the photoelectrons is independent on the intensity of the light source.

(d) For each material there is a threshold frequency f_o such that no matter how intense the light may be, no photoelectrons will be produced if the frequency of the incident radiation is lower than this frequency.

Einstein proposed an explanation for the photoelectric effect. He proposed that light consists of particles known as photons. Each photon has an energy $E = hf$ where h is the Planck's constant. The photoelectric effect can be explained by the following equation

$$hf = W + E_k \qquad (3.3)$$

where W refers to the energy required by an electron to escape from a given material. W is known as the workfunction of the material. When an electron absorbs a photon with energy hf, the difference $hf-W$ will appear as the kinetic energy E_k of the emitted photoelectron. The maximum kinetic energy of the photoelectron is given by

$$\text{Maximum } E_k = hf - W \qquad (3.4)$$

Hence if the energy of the photon is less than W, no photoelectrons will be produced. The threshold frequency is given by $hf_o = W$. More intense light means many more photons, but the energy carried by each photon is the same since it depends only on its frequency. Hence more intense light will only produce more photoelectrons but the maximum kinetic energy of the electrons remains the same. The agreement between the Einstein model and experimental observation justified Einstein's proposal of the energy of the electromagnetic radiation given by $E = hf$. In addition, by treating light as photons, Einstein also introduced the idea of light exhibiting particle-like behaviour.

3.1.4 Wave Particle Duality

Since the introduction of light in the form of photons with particulate nature, scientists began to wonder if matter, considered to be made of particles, might also have a wave nature. Louis de Broglie was the first person to provide an insight into the wave nature of matter. The wavelength λ of a particle, according to de Broglie, is given by

$$\lambda = h/p \tag{3.5}$$

where p is the momentum of the particle. This wave nature of particles was confirmed by the observation of electron diffraction. Once the idea of the wave nature of a particle was established, rapid developments followed that provided a theory to determine the wave properties of a particle moving in the presence of a conservative field.

3.1.5 Heisenberg's Uncertainty Principle

With the wave description, it is impossible to know simultaneously and with exactness both the position and the momentum of a particle. Suppose we know the position, x, of a particle very precisely, then we cannot simultaneous determine the momentum, p, of the particle very precisely. The uncertainty in the position, Δx, and the uncertainty in the momentum, Δp, follows the Heisenberg's Uncertainty Principle, $\Delta x \times \Delta p > h/2\pi$. Any measurement made has to satisfy the uncertainty relation and be of limited precision. The classical concept of having an arbitrarily precise knowledge of both x and p does not apply.

3.2 BASIC POSTULATES OF QUANTUM MECHANICS

Consider a physical system consisting of a particle. Quantum physics proposes a special function known as the *wavefunction* that determines everything that can be known about the system. The wavefunction is a function of position and time, $\psi(r,t)$ and is mathematically a complex function. The product of a wavefunction $\psi(r,t)$ and its complex conjugate $\psi^*(r,t)$ gives $|\psi(r,t)|^2$ that represents the probability density of finding the particle in a particular state. Hence the probability of finding the particle in a volume element dV is given by $|\psi(r,t)|^2$ dV. Since the probability of finding the particle in all space is 1, the wavefunction satisfies the following normalisation condition:

$$\int_{-\infty}^{\infty} |\psi(r,t)|^2 \, dV = 1 \qquad (3.6)$$

Once the wavefunction that describes the system is known, how does one obtain the various physical observables of the system? For example, if we are interested in the energy E of the particle, how can we determine E if we know $\psi(r,t)$? With each physical observable, there is an associated mathematical operator that can be used to "operate" on the wavefunction. The action of the operator is to carry out a mathematical operation on the wavefunction and extract the value of the observable. Mathematically it can be represented as

$$Q\psi = q\psi \qquad (3.7)$$

where Q denotes the operator while q denotes the observable value. For example, if the operator Q chosen is the energy operator, then the value q corresponds to the energy value. For the operator Q, there may exist a special set of functions which are known as the eigenfunctions ψ_j of the operator

$$Q\psi_j = q_j\psi_j \qquad (3.8)$$

with the corresponding eigenvalues q_j. This set of functions form a complete and basic set of linearly independent functions. Any wavefunction representing a physical system can be expressed as a linear combination of the eigenfunctions of any physical observable of the system.

$$\psi = \sum a_j\psi_j \qquad (3.9)$$

where a_j represents a coefficient related to the probability of the particular eigenfunction. Hence the operator Q can be used to extract a linear combination of eigenvalues multiplied by coefficients related to the probability of their being observed.

Once the wavefunction ψ that describes a physical system is known, the expectation value of the physical observable, q, can be expressed in terms of the wavefunction and the operator, Q, associated with the physical observable as follows:

$$\langle q \rangle = \int \psi^* Q \psi dV \tag{3.10}$$

Note that the wavefunction is assumed to be properly normalised and the integration is over all space. If the wavefunction is represented as a linear combination of the eigenfunctions of the operator Q, then the above operation would give rise to the possible values for the physical observables multiplied by a probability coefficient. Hence this is essentially a weighted average of the possible observable values.

For a physical system that is free of external interactions, the evolution of the physical system with time is given by

$$H\psi = i \frac{h}{2\pi} \frac{d\psi}{dt} \tag{3.11}$$

where $i = \sqrt{-1}$ and H is the Hamiltonian operator. This equation is derived from the classical Hamiltonian with the substitution of the classical observables by their corresponding quantum mechanical operators. The role of the Hamiltonian is contained in the Schrödinger equation.

3.2.1 Schrödinger Equation

As we have seen thus far, the wavefunction for a physical system contains everything there is to know about the system. How do we find the exact form of this wavefunction? There are many different varieties of dynamical systems, so how do we find the wavefunction that corresponds to the dynamical problem? Erwin Schrödinger formulated an equation that allows the wavefunction to be determined for any given physical situation. The Schrödinger equation is the analogue of Newton's equation in Classical Mechanics.

The time dependent Schrödinger equation is given as follows

$$-\frac{h^2}{8m\pi^2} \nabla^2 \psi + V\psi = i \frac{h}{2\pi} \frac{d\psi}{dt} \tag{3.12}$$

where ∇^2 represents the Laplacian given, in Cartesian coordinates, by

$$\nabla^2 \psi = \frac{d^2\psi}{dx^2} + \frac{d^2\psi}{dy^2} + \frac{d^2\psi}{dz^2} \qquad (3.13)$$

and V represents the potential energy term.

The time-independent Schrödinger equation is given as follows

$$-\frac{h^2}{8m\pi^2}\nabla^2\psi + V\psi = E\psi \qquad (3.14)$$

where E represents the total energy of the particle.

For a one-dimensional system, Eq. 3.14 becomes

$$-\frac{h^2}{8m\pi^2}\frac{d^2\psi}{dx^2} + V\psi = E\psi \qquad (3.15)$$

Hence, if the potential energy V of a physical system is known, one can make use of Eq. 3.15 to determine the corresponding wavefunction. Thus the functional form for the wavefunction depends on the potential energy V.

3.2.2 Particle in a Potential Box

A simple problem that is commonly discussed and relevant to nanoscience is the case of a particle of mass m, trapped in a potential box. Consider the one-dimensional potential box with a width L as illustrated in Fig. 3.2. Our main task is to find the wavefunctions that would describe the properties of a particle trapped inside such a potential box. In this case, the particle is restricted to move only in the region $0 < x < L$ where the potential energy is equal to zero. In the regions $x < 0$ and $x > L$, the potential energy increases sharply to infinity such that it is impossible to find the particle in these regions. i.e. $\psi(x,t) = 0$ in these regions.

The next task would be to find the wavefunction for the particle in the region $0 < x < L$, where potential energy $V = 0$. Using Eq. 3.15 with $V = 0$, we have

$$-\frac{h^2}{8m\pi^2}\frac{d^2\psi}{dx^2} = E\psi \qquad (3.16)$$

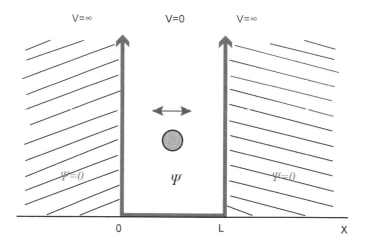

Figure 3.2. One-dimensional potential box.

which can be expressed as

$$\frac{d^2\psi}{dx^2} = -k^2\psi \qquad (3.17)$$

where

$$k^2 = \frac{8m\pi^2 E}{h^2} \qquad (3.18)$$

Equation 3.17 is a typical standing wave equation, and the solutions to the differential equation take the form

$$\psi(x) = e^{ikx} \text{ and } \psi(x) = e^{-ikx} \qquad (3.19)$$

One can easily verify these solutions to the differential equation by direct substitution. Since the particle moves back and forth inside this region $0 < x < L$, we can use a linear combination of the two functions in Eq. (3.19) as the general solution in this case, i.e.

$$\psi(x) = Ae^{ikx} + Be^{-ikx} \qquad (3.20)$$

Note that the wavefunction should satisfy the boundary condition that $\psi(x = 0) = 0$, this leads to the requirement that $B = -A$. Hence we have

$$\psi(x) = A(e^{ikx} - e^{-ikx}) = 2iA\sin(kx) = C\sin(kx) \qquad (3.21)$$

What about the other boundary condition that $\psi(x = L) = 0$? Where does it lead us? We have the following equation

$$\psi(L) = C\sin(kL) = 0 \qquad (3.22)$$

Since C cannot be zero (otherwise we will have no wavefunction), therefore $\sin(kL) = 0$ and this implies $kL = n\pi$ where n is an integer. Substituting this equation back to Eq. (3.18), we have

$$E = \frac{n^2 h^2}{8mL^2} \qquad (3.23)$$

The number n is known as the Quantum Number. Equation (3.23) shows that the energy of the particle trapped in the potential box is discrete and cannot take any arbitrary energy. This situation whereby only certain energy values are allowed is not peculiar to the particle in a box system. It generally holds in any bound physical system, i.e. when a particle is in a bound potential that confines it within a limited region. Such quantization of energy is a common characteristic of nano-physical systems. For a long time, the particle in a box problem remained a quantum mechanics textbook problem. Nowadays, one can readily realise such a potential in a box situation in an artificial quantum well where electrons are confined in a narrow region.

In general, the wavefunction for a particle in a 1D potential box can be expressed as

$$\psi(x) = C\sin(n\pi x/L) \qquad (3.24)$$

In order to determine the expression completely, we make use of the normalisation condition, which requires the probability of finding the particle everywhere to be equal to 1.

$$\int_{-\infty}^{\infty} |\psi(x)|^2\,dx = 1 \qquad (3.25)$$

Since the wavefunction is equal to zero everywhere outside the box, we have

$$\int_{0}^{L} |\psi(x)|^2\,dx = 1 \text{ and so } \int_{0}^{L} C^2\sin^2\left(\frac{n\pi x}{L}\right)dx = 1 \qquad (3.26)$$

i.e. $C = \sqrt{\frac{2}{L}}$ and hence for a particle in a box, the wavefunction is

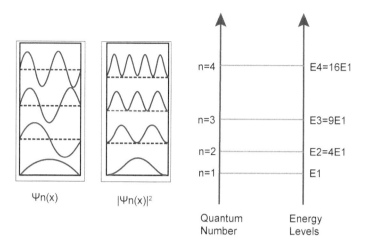

ψn(x) |ψn(x)|²

Figure 3.3. Plots of the wavefunctions $\psi(x)$ corresponding to different quantum states and the square of the wavefunction $|\psi(x)|^2$. The horizontal dash lines in the plot represent the line $\psi(x) = 0$ and $|\psi(x)|^2 = 0$ for these quantum states.

given by

$$\psi(x) = \sqrt{\frac{2}{L}} \sin\left(\frac{n\pi x}{L}\right) \tag{3.27}$$

where $n = 1, 2, 3, 4, 5 \ldots$. Plots of the wavefunctions ψ and the probability density ψ^2 corresponding to the different quantum numbers are shown in Fig. 3.3. We observe discrete quantised energies levels with increasing differences between adjacent levels as the quantum number increases (since $E \propto n^2$).

3.2.3 Generalisation to 3D Potential Box

We can generalise the above discussion to a 3D infinite-wall potential box as shown in Fig. 3.4. In this case, the particle is confined in all three directions inside a box with a dimension of $L_x \times L_y \times L_z$. The potential energy inside the box is $V = 0$ whereas the potential energy outside the box is infinity.

Following a similar discussion to the previous section, we can determine the wavefunction of the particle in the 3D case. Outside

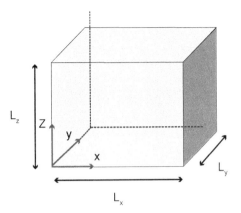

Figure 3.4. Three-dimensional potential box.

the box, $\psi(x, y, z) = 0$, inside the box

$$\psi(x, y, z) = D \sin (k_x x) \sin (k_y y) \sin (k_z z) \tag{3.28}$$

where D is the normalisation constant and

$$k_x = \frac{n_x \pi}{L_x}$$

$$k_y = \frac{n_y \pi}{L_y} \tag{3.29}$$

$$k_z = \frac{n_z \pi}{L_z}$$

Hence the particle is now described by a set of integer quantum numbers (n_x, n_y, n_z). The energy of the particle with mass m is given by

$$E = \frac{h^2}{8m\pi^2} \left[k_x^2 + k_y^2 + k_z^2 \right] = \frac{h^2}{8m} \left[\frac{n_x^2}{L_x^2} + \frac{n_y^2}{L_y^2} + \frac{n_z^2}{L_z^2} \right] \tag{3.30}$$

A few interesting cases follow from the above relations.

Case 1: $L_x = L_y = L_z = L$. Here the energy of the particle simplifies to

$$E = \frac{(n_x^2 + n_y^2 + n_z^2)h^2}{8mL^2} \tag{3.31}$$

Hence the quantum states are defined by each unique combination of the set of quantum numbers (n_x, n_y, n_z). From Eq. (3.31), different permutation of the combination of the quantum numbers give rise to states with the same energy. This is known as degeneracy. For example, $(n_x, n_y, n_z) = (2, 1, 1), (1, 2, 1), (1, 1, 2)$ correspond to states with the same energy value of $6E_o$ where $E_o = h^2/8mL^2$. A plot of the energy levels for the first few values of the energy for the 3D potential box is illustrated in Fig. 3.5(a). The degeneracy for each energy level is also indicated.

The energy difference ΔE between adjacent energy levels depends on the size of the potential box. If L is very small, then we have a potential system where the particle is confined by potential wells in all three dimensions. Such a potential system is known as a *quantum dot*. ΔE is large and typically many times greater than the thermal energy. Hence the physical properties of the quantum dots are strongly influenced by the quantised energy levels and show a sensitive dependence on the size. In nanoscience, size variation is a common strategy to tailor the energy levels of nanosystems.

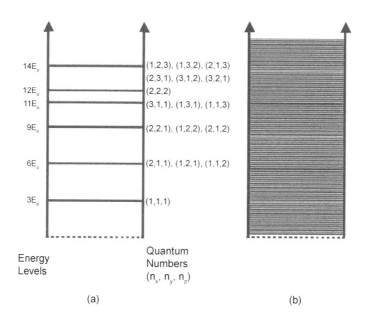

Figure 3.5. (a) Energy levels and quantum numbers for potential box with small dimension. (b) Energy levels for potential box with large dimension.

On the other hand, if L is large, the energy levels will be very close to each other as depicted in Fig. 3.5(b). These closely spaced energy levels practically form a continuous band and it is not practical to account for each of these energy levels. It is more practical to view the system by considering the number of energy levels that can be found in a small energy range. This leads to the idea of density of states and shall be discussed in Chapter 6.

Case 2: $L_x = L_y = L$ and $L_z \gg L_x, L_y$. In such a case, the quantisation condition (3.29) along the z-direction becomes essentially continuous, i.e. there is only a small difference in k_z and energy for n_z and $n_z + 1$. Thus we can write the energy of the particle as

$$E = \frac{h^2}{8m}\left[\frac{n_x^2}{L^2} + \frac{n_y^2}{L^2} + k_z^2\right] \qquad (3.32)$$

where now we have the quantised band characterised by n_x and n_y while k_z is essentially a continuous variable. A plot of the energy values for such as system as a function of k_z is shown in Fig. 3.6.

Such a potential system where the particle is confined by potential wells in two dimensions but free in the third dimension is known as a *quantum wire*.

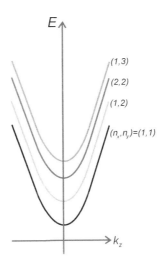

Figure 3.6. Energy versus k_z for quantum wire.

Case 3: $L_y, L_z \gg L_x = L$. In such a case, the quantisation condition (3.29) along both y- and z-directions becomes essentially continuous. Thus we can write the energy of the particle as

$$E = \frac{h^2}{8m} \left[\frac{n_x^2}{L^2} + k_y^2 + k_z^2 \right] \tag{3.33}$$

where the quantised band is characterised by n_x while k_y and k_z are essentially continuous variables. Such a potential system where the particle is confined by potential wells in one dimension but free in the other two dimensions is known as a *quantum well*.

3.2.4 *Potential Barrier Penetration: Potential Step*

So far we have been focusing on potential wells with infinite walls; what happen if the potential energy of the wall is finite? In this case, the wavefunction may extend beyond the classical limits of motion and give rise to an important phenomenon known as potential barrier penetration. As an example, we shall discuss the case for a potential step illustrated in Fig. 3.7.

We divide the system into two regions (I and II). In region I, the particle is free to move around as the potential energy $V = 0$. The Schrödinger equation can be simplified to

$$-\frac{h^2}{8m\pi^2} \frac{d^2 \psi_I}{dx^2} = E\psi_I \tag{3.34}$$

with

$$k^2 = \frac{8m\pi^2 E}{h^2} \tag{3.35}$$

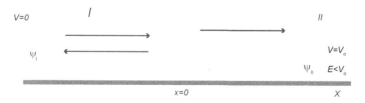

Figure 3.7. Potential step.

Since the particle can travel in the forward as well as backward directions, we can express the wavefunction as

$$\psi_I(x) = Fe^{ikx} + Ge^{-ikx} \tag{3.36}$$

where the first term represents the incident wave while the second term represents the reflected wave. F and G are coefficients that can be determined using the boundary conditions.

For region II, the potential has a finite height of V_o such that $V_o > E$, where E is the energy of the particle. The Schrödinger equation can be expressed as

$$-\frac{\hbar^2}{8m\pi^2}\frac{d^2\psi_{II}}{dx^2} = (E - V_o)\psi_{II} \tag{3.37}$$

We can re-write the above equation as

$$\frac{d^2\psi_{II}}{dx^2} = \kappa^2\psi_{II} \tag{3.38}$$

where $\kappa^2 = 8m\pi^2(Vo - E)/\hbar^2$, and hence we can express the wavefunction in region II as

$$\psi_{II}(x) = He^{-\kappa x} \tag{3.39}$$

To determine the coefficients F, G and H in Eqs. (3.36) and (3.39), $\psi(x)$ and $d\psi(x)/dx$ must be continuous at the boundary points $x = 0$. We have

$$\psi_I(0) = \psi_{II}(0) \tag{3.40}$$

$$\frac{d\psi_I}{dx} = \frac{d\psi_{II}}{dx} \tag{3.41}$$

and thus

$$F + G = H \quad ik(F - G) = -\kappa H$$

For region I

$$\psi_I(x) = F\left(e^{ikx} + \frac{ik + \kappa}{ik - \kappa}e^{-ikx}\right) \tag{3.42}$$

For region II

$$\psi_{II}(x) = F\frac{2ik}{ik - \kappa}e^{-\kappa x} \tag{3.43}$$

Hence we can see that $\psi(x)$ is non-zero inside the potential step and thus it is possible for a particle to penetrate into the potential barrier! This is not allowed in classical physics.

3.2.5 *Potential Barrier and Quantum Tunneling*

As mentioned in the previous section, there is a probability that the wavefunction can penetrate into the potential step. This situation becomes very interesting if the potential step is replaced by a potential barrier. If the potential barrier width W is narrow, it is possible for a particle to penetrate through the potential barrier and appear on the other side! This phenomenon is known as quantum tunneling. Let us consider the potential barrier shown in Fig. 3.8. We divide the system into three region I, II and III as shown.

For regions I and III with $V = 0$, the Schrödinger equation is given by Eq. (3.34) hence we can write down the wavefunction as

$$\psi_I(x) = Pe^{ikx} + Qe^{-ikx} \qquad (3.44)$$

and

$$\psi_{III}(x) = Se^{ikx} \qquad (3.45)$$

We have to account for the presence of the reflected wave in region I while there is no reflected wave in region III.

The intensities of the incident, reflected and transmitted probability current densities, J, are given by

$$J = v\,|P|^2, \quad J = v\,|Q|^2, \quad J = v\,|S|^2 \qquad (3.46)$$

where $v = \frac{\hbar k}{m}$ represents the magnitude of the velocity of the particle. The reflection coefficient R and the transmission coefficient

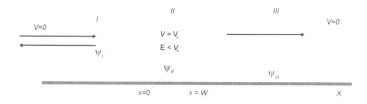

Figure 3.8. Potential barrier.

T are respectively given by

$$R = \frac{|Q|^2}{|P|^2} \tag{3.47}$$

$$T = \frac{|S|^2}{|P|^2} \tag{3.48}$$

Similarly, for region II, the Schrödinger equation is given by Eq. (3.37). The wavefunction is therefore

$$\psi_{II}(x) = Ue^{\kappa x} + Ve^{-\kappa x} \tag{3.49}$$

To determine the coefficients in the wavefunctions, we make use of boundary conditions again.

At $x = 0$,

$$\psi_I(0) = \psi_{II}(0) \text{ and } \frac{d\psi_I}{dx} = \frac{d\psi_{II}}{dx}$$

$$P + Q = U + V \text{ and } ik(P - Q) = \kappa(U - V)$$

At $x = W$,

$$\psi_{II}(W) = \psi_{III}(W) \text{ and } \frac{d\psi_{II}}{dx} = \frac{d\psi_{III}}{dx}$$

$$Se^{ikW} = Ue^{\kappa W} + Ve^{-\kappa W} \text{ and } ikSe^{ikW} = \kappa(Ue^{\kappa W} - Ve^{-\kappa W})$$

which can be simplified to the following equations:

$$\frac{Q}{P} = \frac{(k^2 + \kappa^2)(e^{2\kappa W} - 1)}{e^{2\kappa W}(k + i\kappa)^2 - (k - i\kappa)^2} \tag{3.50}$$

$$\frac{S}{P} = \frac{4ik\kappa e^{-ikW}e^{\kappa W}}{e^{2\kappa W}(k + i\kappa)^2 - (k - i\kappa)^2} \tag{3.51}$$

Thus the reflection coefficient R, (Eq. 3.47) and the transmission coefficient T, (Eq. 3.48) can be expressed as

$$R = \left[1 + \frac{4E(V_o - E)}{V_o^2 \sinh^2(\kappa W)}\right]^{-1} \tag{3.52}$$

$$T = \left[1 + \frac{V_o^2 \sinh^2(\kappa W)}{4E(V_o - E)}\right]^{-1} \tag{3.53}$$

Note that $\sinh(x) = \frac{e^x - e^{-x}}{2}$ and $\cosh(x) = \frac{e^x + e^{-x}}{2}$.

If $\kappa W \gg 1$, then we can use the approximation $\sinh(\kappa W) \sim \frac{1}{2}\exp(\kappa W)$, and the transmission coefficient becomes

$$T \approx \frac{16E(V_o - E)}{V_o^2}e^{-2\kappa W} \qquad (3.54)$$

The probability that the particle can tunnel through the barrier thus depends on the barrier width and the barrier potential height. This relation is an important result for quantum tunneling and the scanning tunneling microscope (see Chapter 8).

3.3 HYDROGEN-LIKE ATOMS: ORBITALS AND ATOMIC STRUCTURES

In this section, we shall discuss the properties of atoms and ions having just one electron. These atoms or ions are known as hydrogen-like atoms. The atom consists of a positively charged nucleus with a charge of $+Ze$ while a single electron (charge $-e$) moves around the nucleus. Here Z corresponds to the number of protons in the nucleus. Assuming that the nucleus behaves like a point charge, the potential energy of such a hydrogen-like atom is given by

$$V = -\frac{Ze^2}{4\pi\varepsilon_o r} \qquad (3.55)$$

where r refers to the separation between the nucleus and the electron, and ε_o is the permittivity of free space. To determine the properties of the electron using quantum mechanics, we are required to solve the Schrödinger equation for the hydrogen-like atom using Eq. (3.55) for the potential energy. Bearing in mind that the electron moves in all three dimensions, thus we have the following Schrödinger equation for hydrogen-like atoms

$$-\frac{h^2}{8m\pi^2}\left(\frac{d^2\psi}{dx^2} + \frac{d^2\psi}{dy^2} + \frac{d^2\psi}{dz^2}\right) - \frac{Ze^2}{4\pi\varepsilon_o r}\psi = E\psi \qquad (3.56)$$

The next task is to solve this equation for the wavefunction and the energy of the system. The solution to Eq. (3.56) is rather complicated, so instead of detailing the complete solution, we shall outline some of the important properties of the equation and its solutions.

Upon solving the above Schrödinger equation, we obtain the following energy equation for the different states of the electron in a hydrogen-like atom:

$$E = -\frac{RhcZ^2}{n^2} \tag{3.57}$$

where R is known as the Rydberg constant ($= 1.0974 \times 10^7$ m^{-1}), and c corresponds to the speed of light. n is the principal quantum number and its value ranges from 1 to ∞. A common form of the equation expressed in units of electron volts is given by

$$E = -\frac{13.6Z^2}{n^2} \text{ (eV)} \tag{3.58}$$

Even though the exact form for the energy differs from the particle in a potential box, the quantisation of energy is a common feature of bound systems where the motion of the particle is restricted. Equation (3.57) applies to hydrogen-like atoms, examples of which include hydrogen ($Z = 1$), deuterium ($Z = 1$), He$^+$ ($Z = 2$), and Li^{2+} ($Z = 3$) (see Fig. 3.9).

How do we know that the energy levels are indeed quantised? The answer lies in the atomic spectra of an atom. When an atom is excited, it will be in one of its excited states; when the atom de-excites, it would go to an energy level with lower energy. The

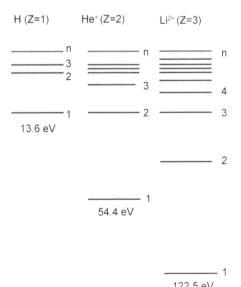

Figure 3.9. Energy levels of H, He$^+$ and Li^{2+}.

difference in energy of the two states is released in the form of radiation, i.e. photons. It is the observations of these emitted photons with specific wavelengths and hence specific energy values that provide experimental evidence of the quantisation of energy. Historically, the hydrogen spectrum was first studied and a series of spectral lines were observed. Some examples are illustrated in Fig. 3.10. The series was identified by the final state n_f, to which more energetic excited states, n_i, make a transition to. Equation (3.59) gives a general relation for the energy of the emitted photon for the transition.

$$E_{\text{photon}} = RhcZ^2 \left(\frac{1}{n_f^2} - \frac{1}{n_i^2} \right) \tag{3.59}$$

In the solution to the Schrödinger equation, three quantum numbers (all integers) are required for each stationary state, corresponding to three independent degrees of freedom for the electron. Besides the principal quantum number n, the other two quantum numbers are the orbital quantum number l, and the z-component orbital quantum number m_l. The values of n range from 1 to ∞, the values of l range from 0 to $n - 1$ and the values of m_l range from $-l$ to l. It can be shown that the following relation

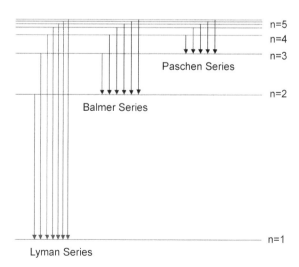

Figure 3.10. Transitions that lead to different spectral series for the hydrogen spectra.

gives the magnitude of the angular momentum, L

$$L = \sqrt{l(l+1)}\frac{h}{2\pi} \qquad (3.60)$$

and the component of the angular momentum along the z-direction L_z is given by

$$L_z = m_l\frac{h}{2\pi} \qquad (3.61)$$

As with the energy, these quantities are also quantised. The quantum number l represents the orbital angular momentum of the electrons, while the quantum number m_l corresponds to its component along the z-direction.

We have learnt about the energy and angular momentum of the electron in the hydrogen-like atom. How about the wavefunction? What about the probability of locating the electron in a certain region near the nucleus of the atom? This can be determined once the wavefunctions that satisfy Eq. (3.56) are determined. In addition, different states of the system are characterised by the set of quantum numbers (n, l, m_l).

To solve for the wavefunction, we make use of the fact that the potential energy (Eq. (3.55)) is spherically symmetric, i.e. depends only on r. Physical problems where the potential energy is only a function of the radial distance r are known as central-force problems. We can simplify the discussion if we re-write the Schrödinger equation using spherical coordinates r, θ, ϕ. The Cartesian coordinate system is transformed to the spherical coordinate system as shown in Fig. 3.11.

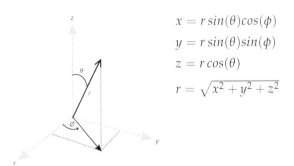

$$x = r\sin(\theta)\cos(\phi)$$
$$y = r\sin(\theta)\sin(\phi)$$
$$z = r\cos(\theta)$$
$$r = \sqrt{x^2 + y^2 + z^2}$$

Figure 3.11. Relationship between spherical coordinates and cartesian coordinates.

Table 3.1 First 10 orbitals and the corresponding quantum numbers of a hydrogen atom.

	n	l	m	s
1s	1	0	0	$1/2, -1/2$
2s	2	0	0	$1/2, -1/2$
2p	2	1	$1, 0, -1$	$1/2, -1/2$
3s	3	0	0	$1/2, -1/2$
3p	3	1	$1, 0, -1$	$1/2, -1/2$
3d	3	2	$2, 1, 0, -1, -2$	$1/2, -1/2$
4s	4	0	0	$1/2, -1/2$
4p	4	1	$1, 0, -1$	$1/2, -1/2$
4d	4	2	$2, 1, 0, -1, -2$	$1/2, -1/2$
4f	4	3	$3, 2, 1, 0, -1, -2, -3$	$1/2, -1/2$

In spherical coordinates, Eq. (3.56) takes the form

$$-\frac{h^2}{8m\pi^2}\left(\frac{1}{r^2}\frac{\partial}{\partial r}\left(r^2\frac{\partial}{\partial r}\right) + \frac{1}{r^2\sin(\theta)}\frac{\partial}{\partial\theta}\left(\sin\theta\frac{\partial}{\partial\theta}\right) + \frac{1}{r^2\sin^2(\theta)}\frac{\partial^2}{\partial\phi^2}\right)\psi$$

$$-\frac{Ze^2}{4\pi\varepsilon_0 r}\psi = E\psi \qquad (3.62)$$

We can re-write the wavefunction as a product of two functions, one that depends only on the distance r and the other one that only depends on the orientation, i.e. θ, ϕ.

$$\psi(r, \theta, \phi) = R(r)Y(\theta, \phi) \qquad (3.63)$$

Here $R(r)$ is known as the Radial Wavefunction and $Y(\theta, \phi)$ is known as the Spherical Harmonic. The radial function depends on the potential energy. On the other hand, the spherical harmonic does not depend on the particular form of the potential energy if the potential energy is only a function of r. The spherical harmonic satisfies the following equation for the quantum operator for the square of the angular momentum, L^2:

|Angular momentum$|^2$ operator

$$L^2Y = \frac{h^2}{4\pi^2}l(l+1)Y \qquad (3.64)$$

Angular momentum along z direction operator

$$L_z Y = m_l \frac{h}{2\pi} Y \tag{3.65}$$

where the complete form for Eq. (3.64) can be written as

$$-\frac{h^2}{4\pi^2} \left[\frac{1}{\sin(\theta)} \frac{\partial}{\partial \theta} \left(\sin\theta \frac{\partial Y}{\partial \theta} \right) + \frac{1}{\sin^2(\theta)} \frac{\partial^2 Y}{\partial \phi^2} \right] = -l(l+1) \frac{h^2}{4\pi^2} Y \tag{3.66}$$

On the other hand, the radial function satisfies the following equation:

$$-\frac{h^2}{8m\pi^2} \left(\frac{\partial^2}{\partial r^2} + \frac{2}{r} \frac{\partial}{\partial r} - \frac{l(l+1)}{r^2} \right) R(r) - \frac{Ze^2}{4\pi\varepsilon_0 r} R(r) = ER(r) \tag{3.67}$$

Once the solutions to the wavefunction are obtained, we can plot $|\psi|^2$ and this gives the probability density distribution; the probability of finding the electron in any region is equal to an integral of the probability density over the region. Depending on the quantum numbers of the electron, we can classify different wavefunction for the electron in different states. States corresponding to different l are in different orbitals. Table 3.1 gives a summary of the various states for the hydrogen atom.

Table 3.2 Mathematical equations for the various spherical harmonic functions.

Y_{l,m_l}	Angular Function
$Y_{0,0}$	$= 1/\sqrt{4\pi}$
$Y_{1,0}$	$= \sqrt{3/4\pi}\cos(\theta)$
$Y_{1,1}$	$= -\sqrt{3/8\pi}\sin(\theta)e^{i\phi}$
$Y_{1,-1}$	$= \sqrt{3/8\pi}\sin(\theta)e^{-i\phi}$
$Y_{2,0}$	$= \frac{1}{2}\sqrt{5/4\pi}(3\cos^2(\theta) - 1)$
$Y_{2,1}$	$= -\sqrt{15/8\pi}\sin(\theta)\cos(\theta)e^{i\phi}$
$Y_{2,-1}$	$= \sqrt{15/8\pi}\sin(\theta)\cos(\theta)e^{-i\phi}$
$Y_{2,2}$	$= \frac{1}{4}\sqrt{15/2\pi}\sin^2(\theta)e^{i2\phi}$
$Y_{2,-2}$	$= \frac{1}{4}\sqrt{15/2\pi}\sin^2(\theta)e^{-i2\phi}$

The corresponding shapes of different orbitals are illustrated in Fig. 3.12. These show the electronics charge density in the region near the nucleus of the atom. Mathematical equations for various spherical harmonic functions are shown in Table 3.2. Similarly, mathematical equations and plots of radial wave functions are shown in Table 3.3.

3.4 SPIN

During the 1920s, another series of experiments found surprising results that resulted in further refinement of quantum mechanics. In these so-called Stern-Gerlach experiments, a beam of atoms (e.g. silver or hydrogen) was sent through a region with non-uniform magnetic field distribution before striking a photographic plate for detection. A schematic of the experiment is illustrated in Fig. 3.13. As shown in Fig. 3.13, it was found that the beam of atoms split into two components that were detected at the photographic plate.

Assuming that we chose z-direction as the direction of maximum non-uniformity of the magnetic field distribution, the net

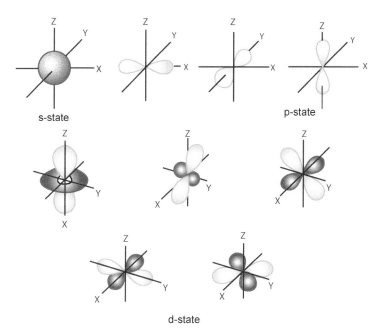

Figure 3.12. Shapes of angular wavefunctions for various states.

Table 3.3 Mathematical equations for the radial wavefunctions and the radial probability distributions.

n	l	R_{nl} $(\rho = 2Zr/na_0)$
1	0	$R_{10} = 2\left(\dfrac{Z}{a_0}\right)^{3/2} e^{-\rho/2}$
2	0	$R_{20} = \dfrac{1}{2\sqrt{2}}\left(\dfrac{Z}{a_0}\right)^{3/2}(2-\rho)e^{-\rho/2}$
	1	$R_{21} = \dfrac{1}{2\sqrt{6}}\left(\dfrac{Z}{a_0}\right)^{3/2}\rho e^{-\rho/2}$
	0	$R_{30} = \dfrac{1}{9\sqrt{3}}\left(\dfrac{Z}{a_0}\right)^{3/2}(6-6\rho+\rho^2)e^{-\rho/2}$
3	1	$R_{31} = \dfrac{1}{9\sqrt{6}}\left(\dfrac{Z}{a_0}\right)^{3/2}\rho(4-\rho)e^{-\rho/2}$
	2	$R_{32} = \dfrac{1}{9\sqrt{30}}\left(\dfrac{Z}{a_0}\right)^{3/2}\rho^2 e^{-\rho/2}$

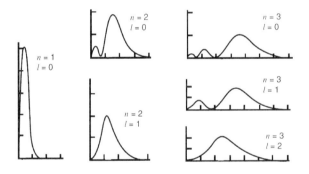

magnetic force on the atoms along the z-direction would be proportional to component of the magnetic moment of the atom in the z-direction. Such a component is proportional to m_l, which has $2l+1$ possible values, i.e. always an odd number. Hence the number of components in the Stern Gerlach experiment is anticipated to be an odd number. Clearly this is different from the experimental observations of only two components. Thus it was proposed that the electron has an intrinsic angular momentum apart from its orbital angular momentum. This intrinsic angular momentum is known as electron spin. Dirac carried out detail analysis of the properties of the electron and confirmed the fundamental nature of the electron spin. He concluded that the electron spin can be described by a new quantum number s which takes the value $\frac{1}{2}$.

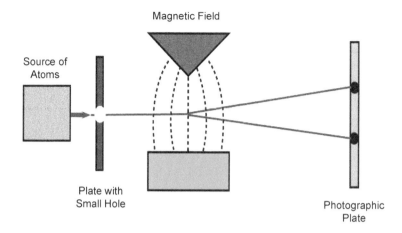

Figure 3.13. Schematic of the Stern Gerlach Experiment.

The magnitude of the spin angular momentum for the electron is given by

$$S = \frac{h}{2\pi}\sqrt{s(s+1)} \tag{3.68}$$

The electron spin is quantized with the z component of the spin angular momentum given by

$$S_z = \frac{h}{2\pi}m_s = \pm\frac{h}{4\pi} \tag{3.69}$$

The two values for S_z correspond to the two components observed in the Stern-Gerlach experiments in the case of silver and hydrogen atoms where there is no contribution from the orbital angular momentum. Hence, the electrons are completely characterized by the set of quantum numbers (n, l, m_l, m_s).

Further Reading

Young and Freedman. *Sears and Zemansky's University Physics with Modern Physics*, 11th Edition (Pearson Addison-Wesley).

Exercises

3.1 An electron is confined within a thin layer of a semiconductor. If the layer can be treated as an infinitely deep one-dimensional potential well, calculate its thickness if the difference in energy between the first ($n = 1$) and second ($n = 2$) levels is 0.05 eV.

3.2 Instead of the 1D potential box as shown in Fig. 3.2, consider a particle moving in the following 1D potential box where the walls are located at $x = -L/2$ and $x = L/2$.

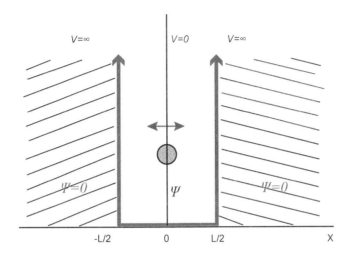

Derive an expression for the energy of the particle
Find the wavefunctions and probability densities for the first five states for the particle.
Draw the wavefunctions and probability densities for the first five states.

3.3 A particle with mass $m = 9.1 \times 10^{-31}$ kg is confined within a one-dimensional potential box with length L and infinite potential on both sides. Starting from the ground state ($n = 1$), among the spectral lines that the particle can absorb at room temperature are two adjacent spectral lines with wavelength 1.0304×10^{-7} m and 5.4953×10^{-8} m. Find the length L.

3.4 An electron collides with a gas of atomic hydrogen, all of which is in the ground state. What is the minimum energy (in eV) the electron must have to cause the hydrogen to emit a Paschen line photon (i.e. transition from higher excited state into the $n = 3$ state)?

3.5 Which of the following is not true regarding the hydrogen atoms?

(i) The energy of the system is quantised.
(ii) The energy difference between one energy level and the next higher energy level decreases with increasing quantum number n.
(iii) The electron is in a confining potential.
(iv) The emission spectra of the hydrogen atom are all in the visible light region.
(v) The properties of the electron are described by a probability wave function.

3.6 An electron with energy of 4.50 eV incident upon a potential barrier of 5.00 eV and thickness L = 950 pm. Calculate the probability that the electron will quantum-mechanically tunnel through the barrier.

This page intentionally left blank

Chapter Four

From Atoms and Molecules to Nanoscale Materials

The discovery of the atom as the basic fundamental unit of matter discussed in Chapter 3 has allowed scientists to see matter as an ensemble of atoms held together by interatomic forces. In this chapter, we will discuss some fundamental concepts and interactions that describe how these ensembles are held together to form solids and materials.

4.1 ATOMS AND THE PERIODIC TABLE OF ELEMENTS

In Section 3.3, the Schrödinger equation was solved for the one-electron hydrogen-like atom and an equation for atomic energy states was obtained:

$$E_n = -\frac{RhcZ^2}{n^2} \tag{3.57}$$

This is a general equation as Z is the atomic number (i.e. number of protons in the nucleus) that uniquely identifies a *chemical element*. In the eighteenth century, scientists such as Dmitri Mendeleev found that when known elements are tabulated according to increasing Z values, the periodicity or trend of chemical properties can be arranged in an orderly pattern into the

Science at the Nanoscale: An Introductory Textbook
by Chin Wee Shong, Sow Chorng Haur & Andrew T S Wee
Copyright © 2010 by Pan Stanford Publishing Pte Ltd
www.panstanford.com
978-981-4241-03-8

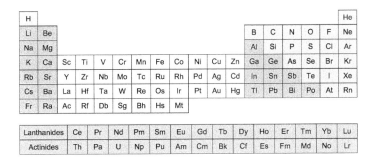

Figure 4.1. A simplified Periodic Table of Elements.

Table 4.1 Orbital representation and the quantum numbers.

Name and Symbol	Allowed Values	Orbital Representation
Principal quantum number (n)	$1, 2, 3, \ldots$	Energy & size of orbital
Angular momentum quantum number (l)	$0, 1, 2, \ldots (n–1)$	Shape of orbital $(s, p, d, f, \ldots)$ and the orbital angular momentum
Magnetic quantum number (m_l)	$0, \pm 1, \pm 2, \ldots, \pm l$	Orientation of the orbital angular momentum (e.g. p_x, p_y, p_z)

Periodic Table of Elements (Fig. 4.1)[1]. In the layout of the Periodic Table, each horizontal period and each vertical group of elements have specific related properties as a consequence of the way their electrons are arranged among the energy states (i.e. their *electron configurations*).

In Chapter 3, we learn that the energy states of electrons are completely characterised by the set of quantum numbers (n, l, m_l, m_s). Chemists often use the concept of atomic orbital (AO) to represent these energy states. Thus, each AO is characterised by a set of three quantum numbers (n, l and m_l). An electron which has either spin up or down and described by the corresponding set (n, l, m_l, m_s) is said to "occupy" that particular AO (Table 4.1).

---

[1] The readers may refer to a Periodic Table put up on the web e.g.: http://www.-webelements.com/

For example, the set ($n = 2, l = 1$ and $m_l = 0, \pm1$) corresponds to the $2p_x$, $2p_y$ and $2p_z$ orbitals (Fig. 3.12). According to Pauli exclusion principle,[2] there can only be six 2p electrons pair-wise ($m_s = \pm1$) in each of these orbitals. Correspondingly, we see that completely filled d orbitals ($l = 2$ and $m_l = 0, \pm1, \pm2$) have 10 electrons (d^{10}) and that of f orbitals ($l = 3$ and $m_l = 0, \pm1, \pm2, \pm3$) have 14 electrons ($f^{14}$). For any n value, there will be an element which has all its corresponding AOs occupied fully with electrons. We called such electron configuration a *closed shell* structure. The elements situated at the right-most column in the Periodic Table constitute the group of noble gases, i.e. He, Ne, Ar, Kr, Xe and Rn. Due to the outermost ns²np⁶ closed shell structure, they exist mainly as monoatomic inert gases in the atmosphere.

In contrast, if we now move to the second column from the right in the Periodic Table, we find the group of halogens: F, Cl, Br, I and At, with electron configuration ns²np⁵. This group of elements is known to have a strong tendency to accept one electron to form anions such as F⁻, Cl⁻, etc. Conversely, elements in the left-most column of the Periodic Table have the tendency to donate one electron to form alkali metal cations such as Li⁺, Na⁺, etc. The element gold (Au, $Z = 79$) has the electron configuration: [Xe]4f¹⁴5d¹⁰6s¹. Silver (Ag, $Z = 47$), which is in the same group as gold, has a similar electronic structure: [Kr]4d¹⁰5s¹. Both these elements form metallic solids that are available in many common forms such as wires, foils, and bars. These materials are good conductors of heat and electricity, and are known for their general inertness to chemicals (although Ag readily reacts with sulfur). We shall see later that these properties are modified when the materials exist in nanometre sizes.

The Periodic Table contains all known 117 elements that have been discovered to date. We have thus seen how different permutations of fundamental entities such as protons and electrons can form vastly different elements that make up all matter surrounding us — e.g. complex life-forming structures such as DNA and RNA are simple combinations of C, N, O and H; modern day IT devices are built mainly using Si. While the atom is the

---

[2] Pauli exclusion principle states that no two fermions (particles with $\frac{1}{2}$ integral spin, e.g. electrons) can occupy the same state. Hence in a single atom, if two electrons have the same (n, l, and m_l) value, then m_s must be different, i.e. the electrons in the same orbital must have opposite spins.

Table 4.2 Types of molecular interactions.

Strong Interactions (Primary bonding) 20–200 kcal mol^{-1}	Weak Interactions (Secondary bonding) 0.1–5 kcal mol^{-1}
Covalent bonding Ionic bonding Metallic bonding	Electrostatic interaction van der Waals forces Dipole-dipole interaction London dispersion forces Hydrogen bonding

fundamental unit of all matter, the properties of matter are controlled principally by the interactions between these units.

4.2 MOLECULES AND MOLECULAR INTERACTIONS

There are many types of interactions that hold atoms together, and they may be broadly classified into primary or secondary bonding on the basis of their strengths (Table 4.2). Covalent bonding involves the sharing of electrons between two atoms. Air consists mainly of N_2 molecules that are formed by two nitrogen atoms covalently bonded together. Molecules are essential entities making up most of our surroundings — e.g. air and water. The forces of attraction that hold a molecule together are referred to as *intramolecular* interactions.

4.2.1 Molecular Orbital Theory

In a simplified manner, intramolecular bonding can be seen as the overlap of AOs of two atoms, resulting in a higher electron density in regions shared by the two nuclei (Fig. 4.2(a)). For example, two p orbitals may overlap either head-on or sideways to produce the σ or π bond respectively (Figs. 4.2(b) and (c)). In order for the orbital overlap to yield effective interaction, the orbitals must approach each other in the right orientation (Fig. 4.2(d)). The strength of the resultant bond depends on the extent of overlap, which in turn is affected by the symmetry and relative energy of the two interacting orbitals.

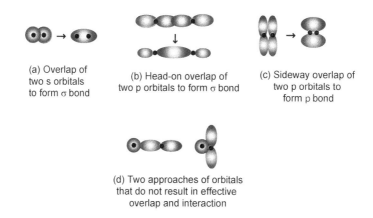

(a) Overlap of two s orbitals to form σ bond

(b) Head-on overlap of two p orbitals to form σ bond

(c) Sideway overlap of two p orbitals to form p bond

(d) Two approaches of orbitals that do not result in effective overlap and interaction

Figure 4.2. Effective overlap between atomic orbitals to form σ and π covalent bonds.

In molecular orbital (MO) theory, molecules are described by MOs in an analogous way as atoms by AOs. A technique known as Linear Combination of Atomic Orbitals (LCAO) is used for constructing MOs. Here, a MO (Ψ) is represented as the summation of i overlapping AOs (ψ_i), each multiplied by a corresponding coefficient (c_i) representing their respective contributions to that MO:

$$\Psi = \sum_i c_i \psi_i \tag{4.1}$$

The coefficients c_i may be determined from the normalization of wavefunctions, similar to Eq. (3.25), and taking into account the overlap of orbitals.

Taking the simplest molecule H_2 as an example, two MOs can be constructed from the two 1s orbitals of the hydrogen atoms (denoted as atom A and B respectively):

$$\Psi(\sigma) = \psi(1s)_A + \psi(1s)_B \tag{4.2}$$

$$\Psi(\sigma^*) = \psi(1s)_A - \psi(1s)_B \tag{4.3}$$

It is noted that while the σ bonding MO corresponds to a higher electron density between the nuclei, the σ^* anti-bonding MO effectively cancels this bonding interaction (Fig. 4.3). The ground state H_2 molecule has two electrons occupying the lower energy σ MO, thus giving rise to a bonding interaction between

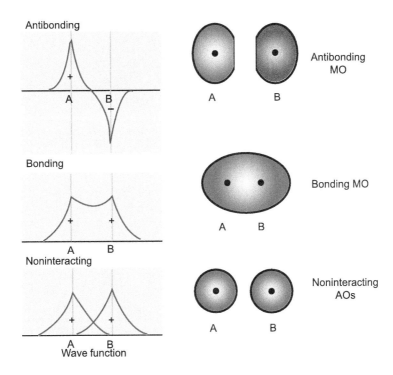

Figure 4.3. Wave function plots for the interacting orbitals of H_A and H_B.

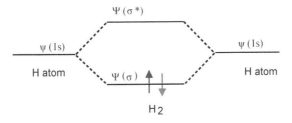

Figure 4.4. Schematic MO energy level diagram for H_2 molecule, with 2 electrons of opposite spins occupying the bonding orbital.

the two H atoms (Fig. 4.4). In the case of He, we know that the He_2 molecule is not stable since the occupancy of σ and σ^* MOs by four electrons cancel out the bonding interaction.

For N_2 molecules, the MO picture must be extended to include 2s and 2p AOs. The schematic MO energy level diagram is given in Fig. 4.5, with the wavefunction notation removed for simplicity.

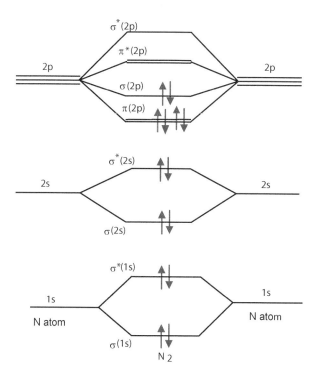

Figure 4.5. Schematic MO energy level diagram for N_2 molecule: the $\pi(2p)$ and $\pi^*(2p)$ orbitals are each doubly degenerate.

The $\pi(2p_x)$ and $\pi(2p_y)$ orbitals are exactly the same energy, i.e. they are *degenerate*. This is similar for the antibonding $\pi^*(2p)$ orbitals. Since the N atoms contribute 7 electrons ($1s^2 2s^2 2p^3$) each, there are altogether 14 electrons to fill into these MOs. The electrons will occupy the MOs from the lowest energy up according to the Aufbau principle[3] and Hund's rule of maximum multiplicity.[4] We can see that there is a total of one σ and two π bonds, making up a triply bonded $N{\equiv}N$ molecule.

---

[3] Aufbau principle postulates the process in which orbitals are progressively "filled" by electrons starting from the lowest available energy states before filling higher states (e.g. 1s before 2s). Generally, orbitals with a lower $(n + l)$ value are filled before those with higher $(n + l)$ values. In the case of equal $(n + l)$ values, the orbital with a lower n value is filled first.

[4] According to Hund's rule of maximum multiplicity, electrons filling into a sub-shell will have parallel spin before the shell is filled with the opposite spin electrons, e.g. two electrons with parallel spin will each occupy $2p_x$ and $2p_y$ orbitals instead of two electrons of opposite spin in one of the 2p orbitals.

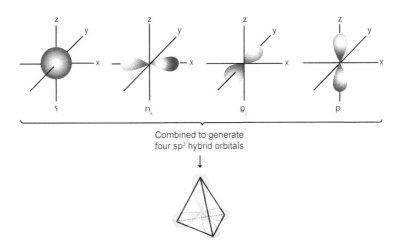

Figure 4.6. Schematic showing the formation of sp^3 hybridised orbitals.

We next consider the MO formation of a simple compound such as methane CH_4. In this case, MO theory predicts that the most favorable interaction occurs only when both the C2s and 2p AOs are involved in bonding with the H 1s orbitals:

$$\Psi(\sigma) = c_H \psi(1s)_H + c_{C(s)}\psi(2s)_C + c_{C(p)}\psi(2p)_C \quad (4.4)$$

Qualitatively, we can view the mixing of one C2s and three C2p AOs as *hybridised* orbitals, hence giving rise to four sp^3 orbitals oriented in a tetrahedral arrangement (Fig. 4.6):

$$\Psi\left(\sigma_{sp^3}\right) = c_H \psi(1s)_H + c_{C(sp^3)}\left\{\psi(2s)_C + c'\psi(2p)_C\right\} \quad (4.5)$$

The relative orientation of these hybridized orbitals thus determines the tetrahedral arrangement of the four C–H bonds in methane. For more complex molecules, appropriate sets of MOs can be iteratively determined using one-electron functions and by minimizing the total energy of the system. The topic of MO calculation is beyond the scope of this introductory text and the readers are referred to the many published MO references for further details.

When two atoms are not similar, e.g. in the H–Cl molecule, the electron distribution is no longer symmetrical but lies more towards the atom with the higher electron affinity. In this case, we say that the Cl atom is more *electronegative* than the H atom and that the HCl molecule is *polar*. In the MOs constructed, the respective coefficients (c_i in Eq. 4.1) will reflect the uneven contributions of the respective AOs. For example, the bonding orbital of HCl may be empirically written as[5]:

$$\Psi\left(\sigma\right) = 0.57\psi\left(1s\right)_H + 0.73\psi\left(2p\right)_{Cl} \tag{4.6}$$

Since the electron probability density is given by the square of the wave function coefficient, we estimate that the bonding electrons spend $\sim 0.73^2 \div \left(0.57^2 + 0.73^2\right) = 62\%$ of their time at the Cl atom. For the extreme situation when the bonding electrons are distributed $\sim 100\%$ over one atom rather than the other, the molecule may be more appropriately described as A^+B^-. The ionic bond thus formed may then be ascribed to the Coulombic force of attraction between the two ions.

4.2.2 Dipole Moment

Classically, two equal but opposite charges $+\delta$ and $-\delta$ separated by a distance l produce a dipole moment μ given as:

$$\mu = \delta \times l \tag{4.7}$$

This is a vector quantity and the direction of the moment is often represented by an arrow $\longmapsto$ as shown in Fig. 4.7. A polar molecule thus possesses a permanent dipole moment due to the unequal electronegativities of its constituent atoms. For example, going down the halogen group $X = F$ to I in the Periodic Table, the dipole moment of the diatomic H–X molecules decreases with the electronegativity of the X atom (Table 4.3).

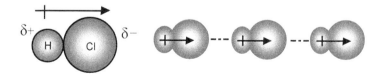

Figure 4.7. Schematics showing polar molecules and the intermolecular interaction between the dipoles.

---

[5] These coefficients are estimated from normalisation taking overlap integral $S \approx \frac{1}{3}$.

Table 4.3 Dipole moments and related properties of some diatomic H–X molecules.

Molecule	H–X Bond Length (pm)	Electronegativity of Element X	Dipole Moment (Debyes*)
HF	91.7	3.90 (reference)	1.86
HCl	127.4	3.15	1.11
HBr	141.4	2.85	0.79
HI	160.9	2.65	0.38

*1 Debye $= 3.336 \times 10^{-30}$ Coulomb meter.

We will see that dipole moment plays an important role in *intermolecular* interactions (Fig. 4.7). Such interactions come into play when molecules are near enough to influence each other. In general, intermolecular interactions can be divided into short-range and long-range forces. Short range forces are those that vary strongly with intermolecular distance, e.g. charge penetration and the Pauli repulsion. They fall off exponentially as a function of distance and are repulsive for interacting closed-shell systems.

Long range energies vary less strongly with distance, and they can be qualitatively understood in terms of classical electrostatic interactions. Long range forces include the electrostatic interaction between two dipoles, between dipoles and induced dipoles, as well as dispersion forces between non-polar molecules. Sometimes, all these long range attractions are known as *van der Waals* forces. We will consider these in more detail in the following sections.

4.2.3 Dipole-dipole Interactions

If two ions (considered as point charges) with charges $z_1 e$ and $z_2 e$ are separated by a distance r, the Coulombic potential developed between them is given by:

$$E_C = \frac{z_1 z_2 e^2}{4\pi\varepsilon_o \varepsilon r} \tag{4.8}$$

Here ε_o is the permittivity of vacuum $= 8.854 \times 10^{-12} \, C^2 N^{-1} m^{-2}$ and ε is the relative permittivity or dielectric constant of the medium between the two ions.

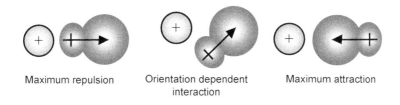

Maximum repulsion   Orientation dependent   Maximum attraction
                                interaction

Figure 4.8. Different orientation of approach between an ion and a polar molecule.

When the ion is placed near a polar molecule, an ion-dipole interaction will occur and the electrostatic potential will depend on the orientation of approach as illustrated in Fig. 4.8. It is found that the attractive orientations, being energetically more favourable, will slightly out-number the repulsive orientations. Rotation of the molecule due to thermal effects, however, randomizes the orientation of the dipoles and the potential thus carries a temperature function in accordance with the Boltzmann distribution.

Extending this picture to the interaction between two permanent dipoles μ_1 and μ_2, the potential energy averaging over all orientations at temperature T is given as:

$$E_C = \frac{\mu_1^2 \mu_2^2}{24\pi^2 \varepsilon_0^2 \varepsilon^2 k T r^6} \tag{4.9}$$

Here, k is the Boltzmann constant. The important point to note here is that the dependence is now on the inverse 6th power of r, and also inversely on T. The latter reflects the situation that at high temperatures, thermal agitation will destroy the mutual interaction between two dipoles. Hence the dipole-dipole interaction becomes important when molecules are condensed into solid, e.g. compressed HCl boils at $-85°$C.

4.2.4 *Induced Dipole Moment*

The presence of an ion or a polar molecule in the vicinity of a second molecule (even if it is non-polar) will have the effect of polarizing the latter. If the electrostatic field strength is E, the induced dipole moment will be αE, where $\alpha =$ the *electric polarizability* of the second molecule. Since the induced dipole moment follows

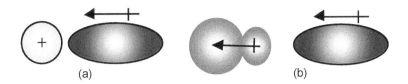

Figure 4.9. Interactions between (a) ion-induced dipole, and (b) dipole-induced dipole.

the direction of the inducing field, it will therefore always interact attractively (Fig. 4.9).

4.2.5 *London Dispersion Forces*

Next, we ask the question of whether there is an interaction between two non-polar molecules as both do not possess a permanent dipole moment. The answer is yes because the electron cloud is always fluctuating and momentary polarization of electron clouds occurs due to instantaneous uneven distributions of electrons. This gives rise to transitory dipoles and attraction between non-polar molecules. Such an attractive intermolecular force is known as the dispersion force, and it is the main attractive force between noble gas atoms in liquefied form.

The theory of dispersion forces was worked out in 1930 on the basis of quantum mechanics by Fritz London.[6] The actual calculation is quite involved and will not be discussed here. We just need to know that it varies proportionally to the polarizability of the molecules and to the inverse 6th power of r. Since larger molecules are intuitively more easily polarized, London dispersion forces become stronger as the molecule becomes larger. This trend is exemplified by the halogen diatomic molecules: F_2 and Cl_2 are gases, Br_2 is liquid, while I_2 is a solid at room temperature. Increasing the amount of surface contact will also enhance the dispersion forces. The dispersion force is the main stabilisation force for self-assembled monolayers formed by molecules with long hydrocarbon chains as discussed in Sections 7.2 and 7.3.

---

[6] R. Eisenschitz and F. London, *Zeitschrift für Physik*, **60**, 491 (1930).

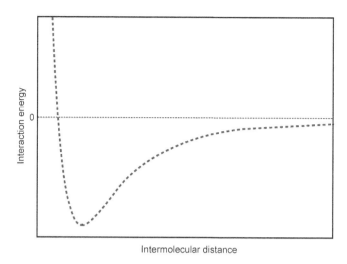

Figure 4.10. Schematic variation of potential energy with intermolecular distance.

4.2.6 *Repulsive Forces and Total Interactions*

The behaviour of condensed phases of matter is determined by a balance between attractive and repulsive forces. We have mentioned earlier that short range forces are repulsive for interacting closed-shell systems. When two molecules are brought too close to each other, their electron clouds will interpenetrate and no longer be able to shield the nuclei. The nucleus and electronic repulsions thus become dominant over the intermolecular attraction. This behaviour at short distances is rather complicated and depends on the nature and electronic structures of the species concerned.

The term van der Waals forces[7] are used to refer to the attractive forces between molecules other than those due to covalent interactions or electrostatic interactions for ions. The sum of long and short range forces gives rise to a minimum, often referred to as the van der Waals minimum (Fig. 4.10). The position and depth of this minimum depends on many parameters. We have learnt that the r dependence of attractive forces is to the inverse

---

[7] In some texts, "van der Waals force" is used to refer solely to the London dispersion force.

6th power; while for many molecules, the repulsive force varies with the inverse 12th power of intermolecular distance. A good approximation to Fig. 4.10 is thus given by the Lennard-Jones potential:

$$PE = -\frac{A}{r^6} + \frac{B}{r^{12}} \qquad (4.10)$$

Here, A and B are both constants to be determined experimentally. The Lennard-Jones potential can also be expressed in various different forms other than that given in Eq. 4.10.

4.2.7 Hydrogen Bonding

Hydrogen bonding is a particular type of dipole-dipole interaction that is important and should be given special consideration. This occurs in molecules containing a hydrogen atom bonded to electronegative atoms such as O, N or F. The dipole generated through this bond will interact with another electronegative atom (known as a *hydrogen bond acceptor*) forming the so-called hydrogen bond (Fig. 4.11). Hydrogen bonding can be either intramolecular or intermolecular. It is among the strongest type of secondary interactions and plays a significant role for molecules such as water, and secondary, tertiary, etc. structures of proteins and nucleic acids.

The simplest example of hydrogen bonding is that between water molecules. As shown in Fig. 4.12, the oxygen atom of the H_2O molecule has two lone pairs of electrons available for

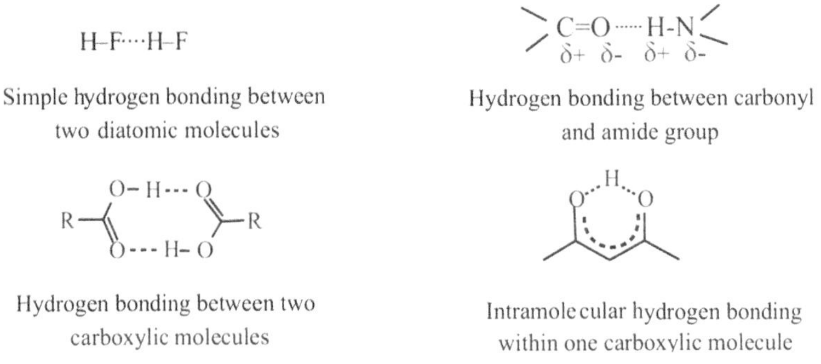

H–F····H–F

Simple hydrogen bonding between
two diatomic molecules

$>$C=O ······ H–N$<$
$\delta+$ $\delta-$ $\delta+$ $\delta-$

Hydrogen bonding between carbonyl
and amide group

O–H ··· O
R—⟨ ⟩—R
O ··· H–O

Hydrogen bonding between two
carboxylic molecules

Intramolecular hydrogen bonding
within one carboxylic molecule

Figure 4.11. Various types of hydrogen bonding.

hydrogen bonding with two other H_2O molecules. This can be repeated so that every molecule is H-bonded to up to four other molecules. In the solid phase, water molecules adopt tetrahedral structures to maximize the number of intermolecular hydrogen bonding among them[8]. When ice melts, some of the hydrogen bonds are broken and water molecules move into the interstitial sites of the partially collapsed structures. This gives water the unique property of having a lower density in the solid state than in liquid state. There is still a large number of hydrogen bonds in liquid water, as evident from its relatively high boiling point compared to some related molecules in Table 4.4.

Hydrogen bonding also plays a significant role in determining the three-dimensional structures adopted by proteins and nuclei acids such as DNA. The natural and precise conformations

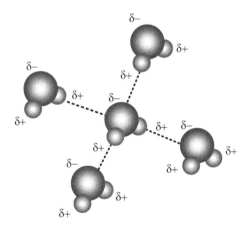

Figure 4.12. Hydrogen bonding in H_2O molecules.

Table 4.4 Properties of water and related substances.

	NH_3	H_2O	HF	H_2S
Melting point (K)	195	273	184	187
Boiling point (K)	240	373	293	212
Dipole moment (D)	1.47	1.85	1.82	0.97

---

[8] Solid state water or ice is known to crystallise into many different structures (or allotropes), the most common one being the hexagonal ice I_h.

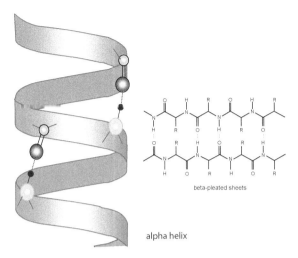

Figure 4.13. Hydrogen bonding giving the secondary structures of protein molecule.

adopted by these biological molecules are crucial for their biochemical and physiological functions. In proteins, the long peptide chains are organized into regular structures known as *alpha-helices* and *beta-pleated sheets*. These are the secondary structures of proteins produced by hydrogen bonding between C=O and N–H groups of the amino acid residues (Fig. 4.13). The double helical structure of DNA, on the other hand, is due largely to hydrogen bonding between the base pairs of the two complementary strands (Fig. 4.14). Such specific shapes of the secondary structures will then facilitate further folding into tertiary and quaternary structures.

In summary, intermolecular interactions are important as they are responsible for many physical properties of materials in solid, liquid, and gaseous phases. In the following section, we examine in more detail the formation and properties of solid materials.

4.3 FROM ATOMS TO SOLID MATERIALS

A piece of bulk solid contains numerous atoms arranged either randomly or in an ordered arrangement, and its properties are given by the average behavior of this collection of atoms. Due to the large number of atoms involved, any theoretical approach

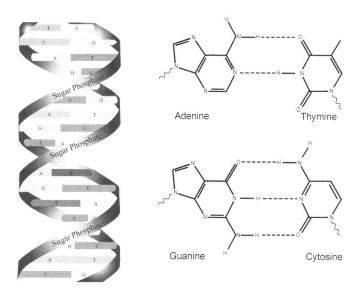

Figure 4.14. Left: The three dimensional structures of DNA double helix. Right: Hydrogen bonding in the Adenine-Thymine (AT) and Guanine-Cytosine (GC) base pairs of DNA.

Table 4.5 Characteristics of different types of solids.

Type	Melting Properties	Examples
Covalent solids	Very high m.p.	Diamond, SiO_2
Ionic solids	High m.p.	NaCl, ZnO
Metallic solids	Moderate to high m.p.	Au, Ag, Fe
Molecular solids	Low m.p.	Ar, CH_4, CO_2

to understand their interactions has to be based on statistical or simulations that extend to infinite arrays. In a more general manner, we may classify the properties of solids according to the type of interatomic or intermolecular bonding between their components as listed in Table 4.5.

4.3.1 Covalent and Molecular Solids

Some materials e.g. diamond, quartz, silicon, germanium, etc., are covalent solids as all atoms in these solids are linked together by

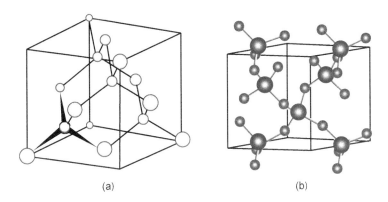

(a)   (b)

Figure 4.15. The structures of some covalent solids: (a) diamond, (b) quartz (SiO_2).

covalent bonds (Fig. 4.15). The characteristics of covalent bonds are largely retained in their solids, e.g. the length of C–C bonds in diamond is nearly the same as those found in molecules such as methane. Hence, covalent solids are typically hard and brittle, and have high melting points.

The covalent bonding, as discussed in Section 4.2, forms only when the overlapping orbitals approach each other in the right orientation. This property of directionality gives an *open* internal structure in covalent solids, with each atom taking up positions at specific directions in space (Fig. 4.15). For example, the tetrahedral bond angle of 109.5° is observed in many cases when sp^3 orbitals are involved. This open structure is unique and contributes to many properties of covalent solids.

Molecular solids, on the other hand, involve weak van der Waals forces. These solids compose of molecules which retain their individual identity, while being held together due to dipole interactions or dispersion forces. Examples are solidified gases such as solid nitrogen, carbon dioxide, etc. Since the binding forces are weak, molecular solids thus have very low boiling points and sublimation temperatures. For polar molecules, some directionality of the internal structure may be observed to optimise the electrostatic or hydrogen bonding interactions between molecules. For non-polar molecules, the London dispersion force is non-directional and hence the molecules tend to adopt close packed structures, similar to metallic or ionic solids as discussed in the following section.

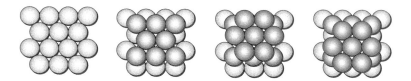

Figure 4.16. Schematic illustration of close-packed structure in solids: (a) closed packed first layer, (b) two layers AB, and two ways for third and subsequent layers: (c) ABABAB...stacking or (d) ABCABC...stacking.

4.3.2 Metallic and Ionic Solids

In general, metallic solids are formed by elements which have a deficiency of electrons. These include elements in the first and second columns of the Periodic Table and most of the transition elements. In metallic solids, the shared electrons are relatively free to move from one atom to another (i.e. they are *"delocalised"*) throughout the solid, thus giving rise to the familiar metallic conductivity and optical properties. Metallic solids tend to adopt close-packed structures in which each atom has 12 surrounding nearest neighbours. This is achieved by having rows of atoms fit into the hollow space in between each row, and with the second layer sitting on top of the hollow spaces in the first layer (Figs. 4.16(a) and (b)). There are two possible ways to stack the third layer, either directly over the first layer (ABAB.. stacking, Fig. 4.16(c)) or in the alternative hollow space of the second layer (ABCABC...stacking, Fig. 4.16(d)). The former gives rise to the hexagonal close packing (hcp) while the latter gives to the cubic close packing (ccp). The ccp structure is equivalent to the face-centred cubic (fcc) structure to be discussed later.

Closed packed structures are the most common form for many metallic solids, e.g. Rh, Ni, Pd, Pt, Cu, Ag, Au, etc. have the ccp structure, while Be, Mg, Ca, Zn, Cd, etc. have the hcp structure. Some metals, e.g. Li, Na, K, V, Cr, Mo, etc., adopt another structure known as the body-centred cubic (bcc) structure that is not closed packed. In the closed packed ccp or hcp structures, the atoms attain the highest packing efficiency of 74% with a coordination number of 12. In a less packed structure such as bcc, the packing efficiency is 68% and the coordination number is 8.

Similarly, ions tend to maximise their coordination with neighbouring ions since Coulombic forces are non-specific, i.e. ions will

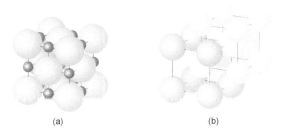

(a)   (b)

Figure 4.17. Schematic structures of (a) NaCl ionic crystal and (b) CsCl ionic crystals; red: Na^+ ions, blue: Cl^- ions, yellow: Cs ions.

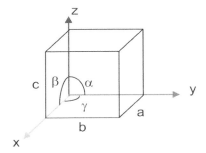

Figure 4.18. Definition of axes, unit cell dimensions and angles for a general unit cell.

interact with all those of opposite charge in the vicinity. In ionic solids such as the rock salt NaCl crystal, the Na^+ and Cl^- ions are arranged in the ccp pattern (Fig. 4.17). Each Na^+ ion is surrounded by six Cl^- ions as coordination >6 will increase the unfavourable repulsion between the negative Cl- ions. For positive ions of larger radii, e.g. Cs^+, we see that more nearest neighbour Cl^- can be accommodated and a bcc structure is adopted instead.

4.3.3 *Crystalline Structures*

As we can see from the previous section, atoms in a solid tend to organized themselves in a periodic pattern to maximize interaction. Such arrangement may be extended over large distances and we say that the solid is crystalline when long range order exists. The internal regularities of crystals may be viewed as the periodic repetition of the sub-units known as the *unit cells*. In three-dimensional space, unit cells are defined by 6 lattice parameters: a, b, c, α, β and γ as shown in Fig. 4.18. A unit cell is the

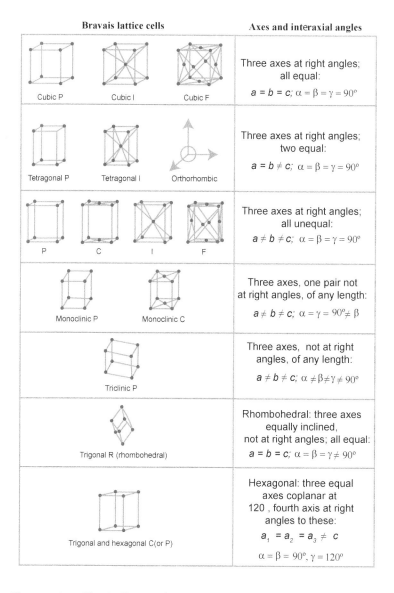

Bravais lattice cells	Axes and interaxial angles
Cubic P Cubic I Cubic F	Three axes at right angles; all equal: $a = b = c;\ \alpha = \beta = \gamma = 90°$
Tetragonal P Tetragonal I Orthorhombic	Three axes at right angles; two equal: $a = b \neq c;\ \alpha = \beta = \gamma = 90°$
P C I F	Three axes at right angles; all unequal: $a \neq b \neq c;\ \alpha = \beta = \gamma = 90°$
Monoclinic P Monoclinic C	Three axes, one pair not at right angles, of any length: $a \neq b \neq c;\ \alpha = \gamma = 90° \neq \beta$
Triclinic P	Three axes, not at right angles, of any length: $a \neq b \neq c;\ \alpha \neq \beta \neq \gamma \neq 90°$
Trigonal R (rhombohedral)	Rhombohedral: three axes equally inclined, not at right angles; all equal: $a = b = c;\ \alpha = \beta = \gamma \neq 90°$
Trigonal and hexagonal C (or P)	Hexagonal: three equal axes coplanar at 120 , fourth axis at right angles to these: $a_1 = a_2 = a_3 \neq c$ $\alpha = \beta = 90°,\ \gamma = 120°$

Figure 4.19. The 14 Bravais lattices. P: primitive cell; I: body-centred cell; F: face-centred cell.

repeat unit that can generate the complete 3D crystal, and is chosen as the sub-unit that has the highest symmetry and smallest volume. There are 14 basic unit cells known as Bravais lattices (Fig. 4.19).

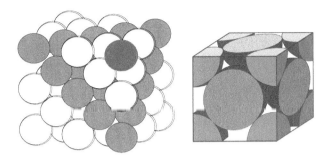

Figure 4.20. The face-centred cubic unit cell of a cubic close-packed (ccp) array.

A full discussion of crystal structures and their symmetry is beyond the scope of this textbook, so we will focus the following discussion on cubic unit cells for simplicity.

We saw earlier that metallic solids such as gold will adopt the ccp structure in order to achieve maximum interaction among its atoms. The unit cell for such a structure belongs to the face-centred cubic cell (Fig. 4.20). In the case of gold, the unit cell dimension a was found to be 4.0786 Å from X-ray diffraction study[9]. Since every unit cell contains four gold atoms (Fig. 4.20), a simple calculation gives $\sim$125 atoms in a gold nanocrystal of $\sim$2 nm. This is a small number compared to that in bulk crystals, and has given rise to the phenomenon of "quantum confinement" discussed in Chapter 6. In the real situation, preparation often produces truncated cubes rather than perfect cubic morphology. In Fig. 4.21, cubic crystal structures adopted by some common solids are given.

4.3.4 Crystal Planes

In the discussion of crystal structures and surface properties (Chapter 5), it is important to identify internal planes that cut through the crystals. For example, the (100) and (111) planes in a cubic crystals are shown in Fig. 4.22. We should realise that these are imaginary planes that pass through the extended crystals and not just end inside the unit cell.

---

[9] L. G. Berry, *Selected Powder Diffraction Data for Minerals*, Joint Committee on Powder Diffraction Standards, Pennsylvania, 1974.

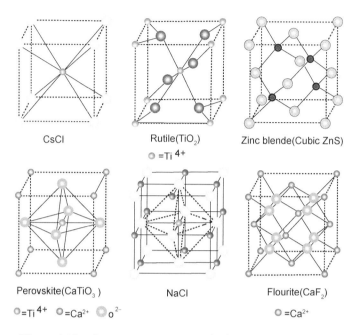

Figure 4.21. Some common types of cubic crystal structures.

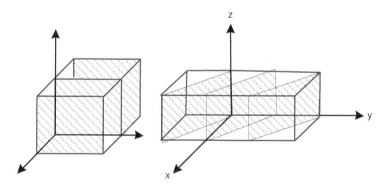

Figure 4.22. The (100) and (110) plane of a cubic unit cell.

The (hkl) notations are *Miller indices*, which are used to denote planes that intercept at positions a/h, a/k, and a/l (or some multiples of these) of the x-, y- and z-axes respectively. Since the indices are defined as the inverses of the intercepts, an index of zero means that the planes are parallel to that particular axis, e.g. the (100) plane is parallel to both the y- and z-axes. While it may not be trivial for gold cubic crystals, we could see clearly that the

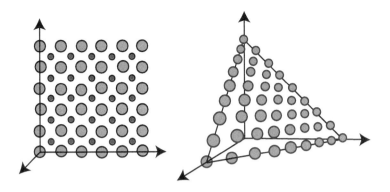

Figure 4.23. Schematic to illustrate the difference between (100) and (111) plane of a NaCl crystal.

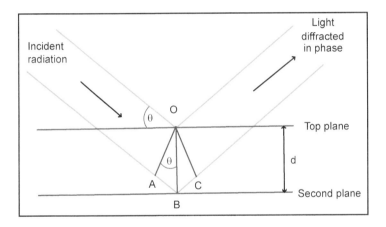

Figure 4.24. Diffraction of x-ray radiation from periodic planes.

two planes (100) and (110) have very different atomic arrangement in the NaCl crystal (Fig. 4.23) and hence are rather different in their chemical behaviours.[10]

The structures of crystals are often analyzed by X-ray diffraction (XRD) technique. When x-ray radiation of wavelength λ passes through the periodic planes of atoms (Fig. 4.24), the diffraction angle θ at which constructive interferences are detected is given

---

[10] In these simple cases, crystal planes correspond to layers of atoms, but this is not generally the case for more complex crystal structures.

by the Bragg's law:

$$\lambda = 2d_{hkl} \sin\theta \qquad (4.11)$$

Here, d_{hkl} is the spacing between the (hkl) planes that diffract the radiation. This is the perpendicular distance between the parallel planes and is given for the cubic crystals as:

$$d_{hkl} = \frac{a}{\sqrt{h^2 + k^2 + l^2}} \qquad (4.12)$$

The XRD patterns of a powdered sample can be used to identify its crystalline phases and structural properties. An observed shift in peak position indicates a change in d-spacing and hence an internal strain in the crystals. An observed peak broadening is also frequently attributed to the finite size of the crystals. The Debye-Scherrer equation relates peak width to the crystallite size D:

$$D = \frac{K\lambda}{W \cos\theta_W} \qquad (4.13)$$

The peak width W is measured as the full width at half maximum of a diffraction peak at θ_W. K is the Scherrer constant of value between $0.89 < K < 1$ and is often taken as unity for common crystals. The broadening of XRD peaks is illustrated in Fig. 4.25 for cubic CdS nanocrystals prepared from solution. It is shown, for the smallest CdS nanocrystals, that peak broadening has caused the diffraction patterns between 40–50 degrees to merge.

4.4 FROM MOLECULES TO SUPRAMOLECULES

We have discussed in Section 4.2 simple molecules such as H_2, N_2, CH_4 and HCl. These are relatively small gaseous molecules with dimensions of a few angstroms (1 Å = 0.1 nm). Common molecules have sizes range from dozens of angstroms to several tens of angstroms.

4.4.1 Macromolecules

It is interesting to note that Nature makes use of some small molecular units as building blocks to form molecules of macroscopic sizes (1 micron = 10^3 nm) for specific functional purposes.

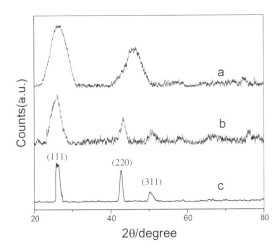

Figure 4.25. XRD patterns for cubic CdS nanocrystals of various average sizes: (a) 2 nm, (b) 3 nm, and (c) 4 nm (from author's lab).

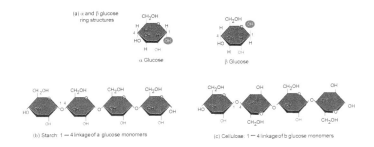

Figure 4.26. α and β glucose molecules and the polymers formed from them.

These may be classified as *macromolecules* and some good examples are the starch and the cellulose, which are formed from either the α- or β-glucose molecules as shown in Fig. 4.26. Such macromolecules composing of repeating structural units connected by covalent bonds is known as *polymer*. The final dimensions of these macromolecules obviously depend on the overall number of repeating units (i.e. the *chain length*) as well as the way of connection (i.e. linear or branching). The process of making such polymer from its components (i.e. the *monomers*) is known as *polymerisation*. Thus, when starch or cellulose molecules are polymerised from glucose monomers, an H_2O molecule is lost for every new C–O covalent linkage forms between the glucose molecules.

Mankind has learnt to mimic nature by producing a variety of macromolecules as synthetic polymers, or plastics. Thus, small molecules containing specific functional groups are chosen and chemically reacted into linear or branching chains of polymers. Common examples are nylon (polyamides) and PVC (polyvinyl chloride) that have taken over traditional raw materials such as fabrics, wood, concrete and clay.

Chemists and material engineers have learnt to control the polymerisation process to optimise the molecular properties (i.e. chain length, branching, density, crystallinity, etc.) in order to obtain the desired material properties. One successful example is polyethylene (PE), which consists of simple $-(CH_2-CH_2)_n-$ repeating units. While high density polyethylene (HDPE, density ≥ 0.941 g/cm^3) is industrially processed to have a low degree of branching and thus has stronger intermolecular forces, low density polyethylene (LDPE, density between 0.910–0.940 g/cm^3) is produced with a high degree of branching and thus the chains do not pack into well crystalline order. Thus HDPE is suitable for applications such as packaging as jugs, bottles and water pipes; whereas molten LDPE has desirable flow properties for it to be processed into flexible plastic bags and film wrap. Our ability to make synthetic polymers and to control their material properties have revolutionised the way of living in today's world.

While scientists and engineers are good at optimising material properties at the macroscopic scale, there is still a long way to learn for making delicate materials or machineries at the nanometre scale. Returning to Nature for a clue, we see that in complex biopolymers such as proteins and DNA (Figs. 4.13–4.14), weak intermolecular interactions are elegantly exploited to organise simple building blocks into functional structures with multi-levels of complexity. On one hand, the strong covalent bonding between the linkages has ensured that the molecule remain intact during reformation; on the other hand, the weak interactions allow the complex structures to be constituted and re-constituted flexibly. Two points are significant for this building blocks approach: (i) information needed for constructing the complex structures are *encoded* already in the building blocks, e.g. the folding of DNA is largely determined by the sequence of base-pairs in the strands, (ii) a large number of weak interactions provides the driving force and the rigidity needed of the complex structures. This brings us to the fascinating research field of supramolecules and self-assembly processes.

4.4.2 Supramolecules

Historically, chemists started off by developing synthetic methodologies that form strong bonding such as the C–C, C–O, etc. covalent bonds. In supramolecular synthesis, attention is focused on the weaker and reversible non-covalent interactions such as hydrogen bonding, van der Waals forces, π–π interaction, metal coordination, etc. These syntheses require careful design of the *motif* (molecules or segments of a molecule) such that it contains the necessary functionalities that will allow it to integrate to form a more complex structure. A simple illustration is given in Fig. 4.27. Here, the –NHC=O motif is incorporated in the molecule *A* to act as both hydrogen-bond donor and acceptor. Two *A* molecules form a cyclic aggregate spontaneously. On the other hand, an incorrectly "coded" molecule *B* will not lead to the ring formation.

One of the key concepts in supramolecular chemistry is *self-assembly*, whereby molecules or segments of molecules integrate

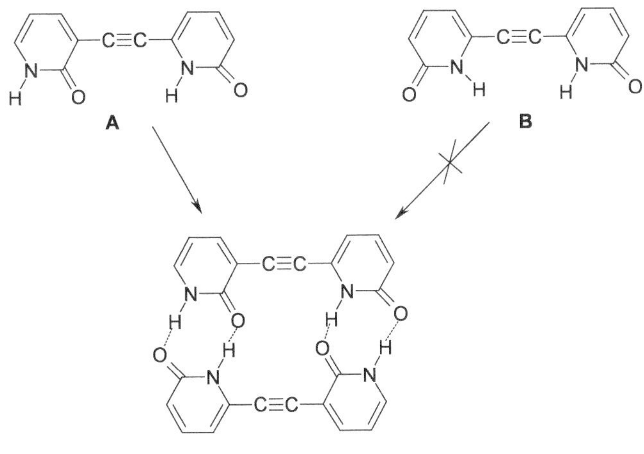

Figure 4.27. Illustration of the –NHC=O motif *encoded* in molecule *A* to form a cyclic assembly, whereas wrongly coded molecule *B* will not.[11]

[11] Y. Ducharme and J. D. Wuest, *J. Org. Chem.* **53**, 5787–5789 (1988).

spontaneously through weak interactions. Molecular self-assembly has allowed the construction of very complex and challenging molecular architectures that would have been too tedious to prepare in a stepwise sequential manner. Self-assembly phenomena are not new and we can look to nature for inspiration. The formation of micelles and monolayers of surfactant molecules (Chapter 5) is an example of self-assembly. The self-assembly technique is also utilized in the bottom-up fabrication of nanomaterials as will be discussed further in Chapter 7.

The preparation and assembly of many supramolecules have been demonstrated in the literature. The French chemist Jean-Marie Lehn is one of the pioneers in this field and was awarded the Nobel Prize[12] in 1987 for his contribution to the synthesis of cryptands and the field of supramolecular chemistry. The reader is referred to the monograph written by Professor Lehn for more details on the design principles of supramolecules.[13] In the following, we will discuss two examples of supramolecules that have attracted much attention for their potential uses as molecular machines (Fig. 4.28).

4.4.3 *Molecular Machines*

Figure 4.28(a) shows a rotaxane-based supramolecule. The molecule may be viewed as an interlocking architecture consisting

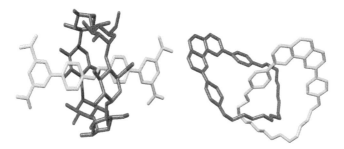

Figure 4.28. Schematic drawings of two supramolecules: (a) an assembly of a rotaxane[14] and (b) an assembly of a catenane.[15]

---

[12] The prize was shared together with D. Cram and C. Pedersen for independent work in the area.

[13] J.-M. Lehn, *Supramolecular Chemistry*, Wiley-VCH (1995)

[14] C. A. Stanier, M. J. O Connell, H. L. Anderson and W. Clegg, *J. Chem. Soc. Chem. Comm.*, 493 (2001).

[15] M. Cesario, C. O. Dietrich-Buchecker, J. Guilhem, C. Pascard and J. P. Sauvage, *J. Chem. Soc. Chem. Comm.* 244 (1985).

of a "dumbbell shaped axle" (blue) and a macrocyclic "wheel" (green). The macrocyclic wheel can rotate around and slide along the axis of the dumbbell. The rotaxane can function as *molecular switch* by controlling the position of the ring from one position to another on the axis.

Another proposed component of molecular machines is the catenane family of which one example is shown in Fig. 4.28(b). This is again an interlocking architecture consisting of two interlocked macrocyclic rings, which are not separable unless one of the covalent bonds is broken. There are many different designs of catenanes, all of them work on the basis that the rings can rotate with respect to each other, with weak interactions between specially encoded motifs on the rings that determine their preferred configurations.

Complex molecular structures have been built in nature to carry out specific physiological functions. An example is *ATP synthase*, which is an enzyme that synthesises adenosine triphosphate (*ATP*) from adenosine diphosphate (*ADP*):

$$\text{ADP} + \text{Phosphate} \xrightarrow{\text{ATP Synthase, } H^+} \text{ATP}$$

Energy is required for this reaction and this is often driven by protons moving down an electrochemical gradient. The enzyme has a large mushroom-shaped structure ~ 10 nm across and ~ 8 nm high, consisting of two segments F_0 and F_1 (Fig. 4.29). The hydrophobic F_0 segment is embedded in the membrane and performs proton translocation, while the hydrophillic F_1 segment protrudes into the aqueous phase to perform ATP synthesis. During the reaction, conformational changes in some segments of the enzyme generate a rotation, making ATP synthase the smallest rotary machine known in nature.[16]

In summary, scientists are looking into the nano-world of biological molecules for inspiration in designing molecular electronic and mechanical machines. One major question is how to provide suitable energy inputs to drive these nano-machines. While temperature gradient is difficult to maintain over such small dimensions, chemical reactions will produce side-products that need to be transported away efficiently. One promising solution for powering man-made nano-machines is the use of

---

[16] P. D. Boyer, J. E. Walker and J. C. Skou shared the 1997 Nobel Prize in Chemistry for their independent work on ATP synthase and the other ion-transporting enzyme, Na$^+$, K$^+$-ATPase.

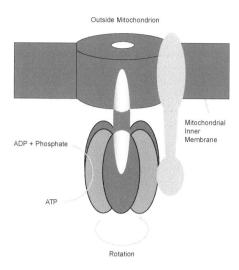

Figure 4.29. The rotation of ATP synthase.

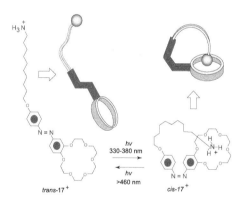

Figure 4.30. Schematics showing motion of the cationic tail which can only interact with its macrocyclic head in the *cis* isomer; the *cis-trans* isomerisation is induced by radiation. [V. Balzani, A. Credi, F. M. Raymo and J. F. Stoddart, Artificial Molecular Machines, *Angew. Chemie. Int. Ed.* 2000, **39**(19), 3348–3391 Copyright Wiley-VCH Verlag GmbH & Co KGGA. Reproduced with Permission.]

radiation. Radiation of precise wavelengths have been conveniently used to induce conformational changes, e.g. in the *cis-trans* isomerisation of a supramolecule shown in Fig. 4.30. In this example, the "lock" and "unlock" positions of the ionic tail group into the ring-shaped head group can be reversibly controlled using radiation at different wavelengths.

Further Reading

M. Ladd, *Chemical Bonding in Solids and Fluids* (Ellis Horwood, 1994).

P. A. Cox, *The Electronic Structure and Chemistry of Solids* (Oxford University Press, 1991).

L. Smart and E. Moore, *Solid State Chemistry — An Introduction* (Chapman & Hall, 1992).

V. Balzani, A. Credi, F. M. Raymo and J. F. Stoddart, *Artificial Molecular Machines, Angew. Chem. Int. Ed.* **39**, 3348–3391 (2000).

Exercises

4.1 On the basis of their electronic structures, answer the following questions for the first row of elements in the Periodic Table: (i) Li forms Li^+ ions readily in solids and solutions, why does Be not form Be^{2+} ions in solids? (ii) Why is there a discontinuity for the trend of variation for ionization energy from N to O? (iii) How do the electron negativities vary from Li to F atom and why? (iv) Which of the elements in (iii) have tendency to form anions?

4.2 The bond distance of HCl molecule is 1.27 angstrom. Determine what portion of a unit electron charge is transferred from hydrogen atom to chlorine atom.

4.3 In addition to hydrogen bonding, there are at least two other interactions that are important for DNA double helix (Fig. 4.14). Identify these and give a brief description of each.

4.4 The coordination number of an ion A in ionic solid AB is determined by the ratio of their radius r_A/r_B. Work out the maximum coordination number possible for A if the ratio is 0.3.

4.5 Cis-trans isomerisation such as that shown in Figure 4.30 requires energy input. Convert these energies into the usual unit of kJ mol^{-1}.

Chapter Five

Surfaces at the Nanoscale

Surfaces play an important role in many aspects and applications of materials. Specifically, the surface of a material object refers to its outermost layer of atoms or molecules that comes into direct contact to its surroundings. In general however, surface properties often involve the subsequent one or few more layers of atoms beneath due to surface reconstruction or reorganisation.

Fundamentally, surface properties are not altered when materials are reduced to the nanoscale. Nonetheless, surface behaviour and reactivity become more significant at the nanoscale and thus warrant special consideration in this chapter. In the last section of this chapter, we will also introduce *microemulsions* and *surfactants*, which are often encountered in the chemical synthesis of nano-structures.

5.1 SURFACE ENERGY

5.1.1 *Fraction of Surface Atoms*

One of the main changes when materials are reduced to the nanoscale is the tremendous increase in the fraction of atoms that reside on the surface in comparison to the total number of atoms in the material. As a simple illustration, we compare two sizes of gold fcc crystals in cubic form as shown in Fig. 5.1. It can be estimated quite easily from the packing efficiency (Chapter 4) that there is a total of $\sim 5.9 \times 10^{22}$ closely packed atoms in a 1 cm^3 gold cube. In a cube of this size, only $\sim 2 \times 10^{-6}$ % of the atoms are residing on the six facets of the cube. Hence, any slight defect

Science at the Nanoscale: An Introductory Textbook
by Chin Wee Shong, Sow Chorng Haur & Andrew T S Wee
Copyright © 2010 by Pan Stanford Publishing Pte Ltd
www.panstanford.com
978-981-4241-03-8

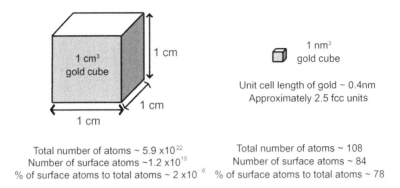

Total number of atoms ~ 5.9 x10^{22} Total number of atoms ~ 108
Number of surface atoms ~1.2 x10^{15} Number of surface atoms ~ 84
% of surface atoms to total atoms ~ 2 x10^{6} % of surface atoms to total atoms ~ 78

Figure 5.1. A comparison of the percentage of surface atoms in (a) 1 cm^3 gold cube to (b) 1 nm^3 gold nanocube.

e.g. a missing atom on the surface, will have insignificant effect on the overall properties of this 1 cm^3 gold cube.

On the other hand, inside a 1 nm^3 gold cube, it can be estimated that there are only about two and a half arrays of fcc atoms or a total of 108 atoms. Among these atoms, ~84 of them or ~78% of the total are surface atoms. It is hence not difficult to expect that the properties of this nanocube are essentially controlled by its surfaces. For even smaller crystals, the percentage of surface atoms may well be approaching 100% (i.e. all atoms are surface atoms!). Hence when materials are reduced to the nanoscale, a dramatic proportion of its atoms become surface atoms. The chemical and physical properties of these materials will therefore be strongly controlled by the behaviour of their surfaces.

5.1.2 *Surface Atoms and Their Energy*

Unlike atoms in the interior of a bulk solid that are fully coordinated chemically, atoms on surfaces have a lower coordination number as they have fewer nearest neighbours. Consequently, these surface atoms have a higher energy than those in the interior, and we define this extra energy as the *surface energy*.

Surface energy is often denoted by the symbol γ, and may be defined as the free energy required to create a unit area of "new" surface of the solid material:

$$\gamma = \left(\frac{\delta G}{\delta A} \right)_{T,P} \tag{5.1}$$

G = Gibbs free energy

A = Surface area

Apparently, some bonds have been broken in creating this new surface. Extensive model calculations using such "broken-bond" models have been performed for a variety of crystal planes and structures[1]

We have learnt in Section 4.3 that atoms are arranged differently on different crystal planes. In the following, we use a simplified version of the "broken-bond model" to illustrate that different crystal planes have different surface energies. By ignoring the interactions of higher order neighbours, we estimate γ as half of ε, the bond strength, multiplying with the number of broken bonds (N_b):

$$\gamma = \frac{1}{2} N_b \varepsilon \rho_a \tag{5.2}$$

ρ_a = number of atoms per unit area on the new surface

This model is obviously oversimplified since it assumes that the bond strength is the same throughout the whole crystal structure, and is similar for both surface and bulk atoms. The former may be true in the case of elemental metallic crystals composed of just one type of atoms. In a simple picture, let us consider cutting away one unit cell along the {100} facet of a fcc metallic solid (Fig. 5.2). Since each atom on the "new" {100} surface is left with 8 coordinating neighbours, it is clear that four bonds have been broken and $\gamma = (4\varepsilon)/a^2$ according to Eq. 5.2.

A similar schematic can be drawn for the {111} surface to show that three bonds are broken and $\gamma = (2\varepsilon\sqrt{3})/a^2$. Thus, this simplified model allows us to see that different crystal facets possess different surface energies. Broken-bond model calculations that take into account second and higher order interactions predict that $\gamma_{(111)} < \gamma_{(100)} < \gamma_{(110)}$ for fcc metals.[1] Generally, it was found that crystal surfaces with lower Miller indices have a lower surface energy than those with higher Miller indices. Commonly

---

[1] S. G. Wang, E. K. Tian and C. W. Lung, *J. Phys. and Chem. Solids*, **61**, 1295 (2000).

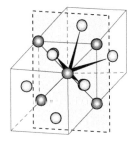

The surface atom on the {100} facet would have four bonds broken,

$$\gamma = \frac{1}{2} \times 4 \times \varepsilon \times \frac{2}{a^2}$$

$$\gamma = \frac{4\varepsilon}{a^2}$$

Figure 5.2. Schematic of a "new" {100} surface, showing four broken bonds for the red coloured surface atom. The broken bonds are illustrated by the green coloured balls in the front unit cell.

Cube {100}   Dodecahedron{110}   Octahedron {111}

Figure 5.3. Forms or shapes of crystals are determined by the surrounding facets.

observed crystals thus often have simple shapes bounded by low index surfaces.

Thermodynamically, the equilibrium shape of a crystal is determined by considering the surface energies of all facets. Typically, a crystal surrounded by {100} facets will adopt the form of a cube, whereas one with {111} or {110} facets will have the form of an octahedron or dodecahedron (Fig. 5.3). This general guideline, however, may not be applied to nano-sized crystals that are grown through kinetically-controlled routes.

5.1.3 *Lowering of Surface Energy*

We often observe that liquid water forms spherical droplets on a substrate. This is a spontaneous behavior as the spherical shape minimizes the total surface area and hence the free energy. Thermodynamically, a material object is stable when in a state of the lowest Gibbs free energy. In order to lower the free energy, some surface processes such as surface relaxation or reconstruction will occur naturally.

Since surface atoms have "unsatisfied bonds", they experience inward and lateral pulls due to their unbalanced coordination (Fig. 5.4(a)). In surface relaxation, atoms in the surface layer may shift inwardly (Fig. 5.4(b)) or laterally (Fig. 5.4(c)) in order to counter these unbalanced forces. In both cases, there is no resultant change in the periodicity parallel to the surface, or to the symmetry of the surface.

In some surfaces, nevertheless, more disruption to the periodicity or symmetry is observed. The broken or dangling bonds of the surface atoms may combine to form strained bonds between themselves (Fig. 5.5). This causes the surface atoms, and sometimes also one or more subsequent layers of atoms below, to be distorted from their equilibrium bulk positions. The surface layer is thus *restructured* with different bond lengths and/or angles. This phenomenon is called surface reconstruction. A well-known example of such is the Si(111)-(7 × 7) reconstruction, which

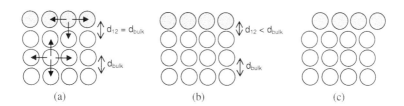

Figure 5.4. Schematic diagram showing (a) surface atoms experience an inward and lateral pull; (b) surface relaxation, the surface atoms shift inwardly; (c) surface relaxation, the surface atoms shift laterally.

Figure 5.5. Surface restructuring or reconstruction, illustrated by the formation of strained bonds between the surface atoms.

has been observed directly with Scanning Tunneling Microscopy (STM), and is discussed in Section 8.3.1.

Surface adsorption is another common way of reducing the surface energy. In this case, a foreign species sticks onto the surface (i.e. *adsorb*), forming bonds or just weak electrostatic or van der Waals interactions with the surface atoms. A good example is the adsorption of hydroxyl groups (-OH) at the surface dangling bonds of silicon wafers after treatment with Piranha solution,[2] thus making the surface hydrophilic. In Chapter 7, we will see that surface adsorption using specific molecules has been vigilantly applied by chemists in the size-controlled synthesis and isolation of nanoparticles from solutions.

During the preparation or processing of nanostructures, there are several dynamic mechanisms that can occur to reduce the overall surface energy of the system. In the most common situation, several nano-sized objects will associate together through chemical or physical attraction at the interfaces. This agglomeration into larger associations or clusters does not alter the individual properties of the nanostructures. It will, however, give rise to difficulty in re-dispersing the associated clusters in solutions. Due to the huge surface energy incurred, attempts to prepare nanostructures without appropriate stabilisation measures are very likely to result in agglomerate formation. Hence, effectively preventing agglomeration is one of the main considerations in the preparation and handling of nanomaterials.

Individual nanostructures will merge into larger structures in order to reduce the overall surface area in two other mechanisms, *sintering* and *Ostwald ripening*. Sintering often occurs at elevated temperatures, during which atoms at surfaces or grain boundaries undergo solid-state diffusion, evaporation-condensation or dissolution-precipitation processes. The individual nanostructures thus change their shapes when they combine with each other, and this often results in a polycrystalline material (Fig. 5.6(a)). Such a process has in fact been advantageously used in the ceramic and powder metallurgy industries.[3]

Ostwald ripening, on the other hand, will eventually produce a single uniform structure with the larger nanostructures growing

---

[2] Piranha is a mixture of sulphuric acid and hydrogen peroxide, used as a common etchant in the microelectronics industry.

[3] J. S. Reed, *Introduction to Principles of Ceramic Processing*, Wiley, New York, 1988.

(a)   (b)

Figure 5.6.   Schematics showing (a) sintering, and (b) Ostwald ripening.

at the expense of the smaller ones (Fig. 5.6(b)). In this context, the Gibbs-Thompson equation used in classical crystallisation theory[4] provides us with the correlation:

$$S_r = S_b \exp \left(\frac{2\sigma V_m}{rRT} \right) \tag{5.3}$$

Here, r is the radius of the crystal, σ is the specific surface energy, V_m is the molar volume of the material, S_r and S_b are respectively the solubility of the bulk crystal and a crystal with radius r. R and T are thermodynamic parameters: R being the gas constant and T is the absolute temperature.

Equation 5.3 suggests that the solubility of a given crystal is inversely dependent on its size. When two nanoparticles of different sizes (say r_1 and r_2, where $r_1 \gg r_2$) are put together in solution, each particle will develop an equilibrium solubility with the surrounding solvent. Thus, the particle with r_2 may dissolve due to its higher solubility and a solute gradient develops. Consequently, a net diffusion of solute from the vicinity of the smaller particle to that of the larger particle occurs. In order to maintain the equilibrium, solute will deposit onto the larger particle while continuing to dissolve from the smaller particle. Such dissolution and condensation processes will continue until the complete dissolution of the smaller particle. Finally, a larger uniform particle is obtained as shown in Fig. 5.6(b).

This Ostwald ripening phenomenon is important especially for solution growth or crystallisation of nanoparticles. In particular, the mechanism results in the elimination of smaller particles and thus the size distribution becomes narrower. Ostwald ripening can be optimised by varying the process temperature and/or by changing the concentration or the solute supply. The process has been advantageously used to prepare nanoparticles of narrow size distributions as discussed in Section 7.2.

---

[4] J. W. Mullin, *Crystallization*, 3rd Edition, Oxford, 1997.

5.2 SURFACE REACTIVITY AND CATALYSIS

In chemical reactions involving a solid material, the surface area to volume ratio plays an important role in the reactivity. This is analogous to the situation where finely crushed ice melts faster than ice cubes. Materials with higher surface area are expected to react more readily because more surface sites are available to react. A famous historical example is the destructive explosion caused by a spark and flour dust in the "Great Mill Disaster" accident in 1878;[5] while grain is not typically flammable, grain dust becomes explosive due to its extremely high surface energy. Thermodynamically, a high surface area provides a strong "driving force" that speeds up processes in the quest to minimise free energy.

High surface area can be achieved either by using materials of very small sizes or materials that possess highly porous structures. In the latter, microporous materials such as zeolites[6] have played an important role in heterogeneous catalysis. A catalyst speeds up reactions by providing an alternative reaction pathway of lower activation energy (E_a) for the system concerned (Fig. 5.7). In heterogeneous catalysis, the catalysts are often solid materials that provide a surface on which the reactant molecule (either in gas or liquid phase) temporally adsorbs. This catalyst surface possesses some active sites such that the adsorbed molecule can reorganize into a form that will facilitate the reaction. This is often followed by fragmentation and desorption of the products or by-products.

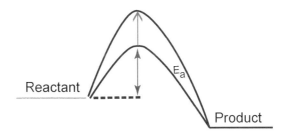

Figure 5.7. Energy diagram showing the effect of a catalyst.

---

[5] See "Washburn 'A' Mill" in Wikipedia: http://en.wikipedia.org/wiki/Washburn_%22A%22_Mill.

[6] Zeolites — a class of hydrated aluminosilicates that contains a highly porous (pore sizes ∼0.3–10 nm) structures.

In general, we may divide the catalytic surface reaction into several consecutive steps: (1) diffusion of reactants to the surface; (2) adsorption of reactants at the surface; (3) chemical transformation of the adsorbed molecule, (4) reaction on the surface; (4) desorption of products from the surface; (5) diffusion of products away from the surface. Suppose any of these steps has a much lower rate constant than all the others, it will become the rate determining step and control the overall rate of reaction. Thus, while higher surface area causes smaller particles to react faster, the situation is less straight-forward for heterogeneous solid catalytic reactions.

Among the many nano-sized materials, supported metallic particles have been most successfully applied as catalysts. One practical example is the use of nano-sized platinum and rhodium particles in the catalytic convertor of car exhaust (Fig. 5.8) to improve its efficiency.

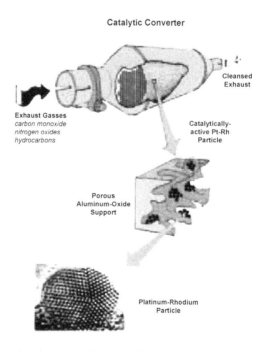

Figure 5.8. A schematic showing the catalytic convertor used in car exhaust. [With kind permission from Springer Science Business Media: *J. Phys. D*, Atomic Resolution electron microscopy of small metal clusters, **19**, 293 (1991), J.-O. Boyin and J.-O. Malm. Copyright © 1991, Springer Berlin/Heidelberg.]

Masatake Haruta first reported the catalytic activity of gold nanoparticles towards oxidation and reduction of hydrocarbons.[7] Gold is traditionally considered an inert metal and catalytically inactive, however the increase in its catalytic activity has been attributed to various factors, e.g. the low-coordination sites on its surfaces, the higher mobility of surface atoms, the higher electronegativity and higher oxidation potential. The relatively higher surface mobility of corner and edge atoms at room temperature is supported by the lowering of its melting point at the nanoscale — gold nanoparticles of 2.5 nm in diameter have been found to melt at $\sim$600°C compared to 1320°C for the bulk gold. In addition, the catalytic activity of gold nanoparticles towards CO reduction has been reported to be optimum at the diameter of $\sim$3.5 nm, when the metal to non-metal transition is observed. In a recent report, nano-sized gold particles of 2–15 nm diameters have been shown to demonstrate higher catalytic activity for many selective hydrocarbon oxidation reactions that are used to make compounds contained in agrochemicals and pharmaceuticals.[8] For an overview of this exciting and growing field of nanocatalysis, the readers may refer to a recent reference on the subject.[9]

5.3 SURFACE STABILISATION

Due to the large energy associated with their high surface areas, nano-sized objects can be considered to be thermodynamically unstable (or *"metastable"*) as there is a natural drive towards reduction of free energy via processes such as agglomeration etc. How was it possible then, as mentioned in Chapter 1, for gold nanoparticles to be prepared in ancient times and used in beautiful stained glass windows in medieval churches?

Since the early days of alchemists in the 17th century, it was already known that adding certain salts or chemicals allows stable *colloids* to be prepared. A *colloid* refers to a suspension of fine particles in liquid phase such as water or organic solvents. Thus it is a two-phase system and a resultant interfacial potential develops.

---

[7] M. Haruta, *Catalysis Today* **36**, 153 (1997).
[8] M. D. Hughes, Y. Xu, P. Jenkins, P. McMorn, P. Landon, D. I. Enache, A. F. Carley, G. A. Attard, G. J. Hutchings1, F. King, E. H. Stitt, P. Johnston, K. Griffin and C. J. Kiely, *Nature* **437**, 1132 (2005).
[9] U. Heiz and U. Landman, *Nanocatalysis*, Springer, 2007.

Michael Faraday called this system *"sols"*, and demonstrated in his lecture the vivid ruby red colour of the gold sols he prepared.[10] Indeed Faraday stated in this historical paper that *"known phenomena seemed to indicate that a mere variation in the size of particles gave rise to a variety of resultant colours"*. This is probably the earliest scientific expression of the "size effect" referred to in Chapter 3.

5.3.1 The Electrical Double Layer

Colloids are unstable thermodynamically. The existence of stable colloids is attributed to electrostatic stabilisation arising from the formation of an electrical double layer. Generally, a net surface charge will develop on the particles through various mechanisms. Due to the presence of this charge on the surface, ions of opposite charge will tend to cluster nearby the particle to form an ionic atmosphere. Two regions of charge are identified — first, the fairly immobile layer of ions that adhere strongly to the particle surface (the Stern layer); second, a diffuse layer of oppositely charged mobile ions (the Guoy layer) that are attracted to this first layer. These inner and outer layers of inhomogeneously distributed charges thus constitute the "electrical double layer" surrounding the colloidal particles (Fig. 5.9).

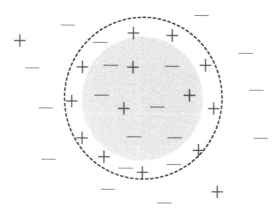

Figure 5.9. A schematic showing the electrical double layer.

---

[10] M. Faraday, *Philosophical Transactions of the Royal Society of London*, **147**, 145 (1847). See also Michael Faraday's lecture slide of the gold sols in the Whipple Museum of the History of Science: http://www.hps.cam.ac.uk/whipple/explore/microscopes/faradaysslide/

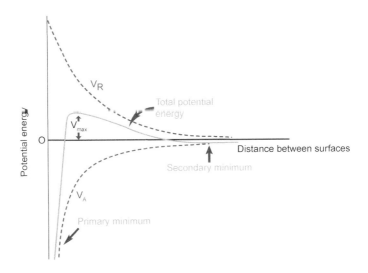

Figure 5.10. A schematic plot of DLVO potential as a function of distance between the surfaces of two particles.

When nano-sized particles are dispersed in a solution, Brownian motion ensures that the particles will move about, constantly colliding with each other. When two particles move close to each other and the two electrical double layers overlap, a repulsive electrostatic force develops. For stability of the dispersion, DLVO theory[11] assumes that there is a balance between the repulsive interactions (V_R) between the double layers on neighboring particles and the attractive interactions (V_A) arising from van der Waals forces between the molecules in the particles. A plot showing the effect of these two opposite potentials, expressed as a function of distance between the surfaces, is shown in Fig. 5.10. Thus, a potential maximum located near the surface is found which is known as the repulsive barrier. Coalescence of two colliding particles will occur only when the collision is sufficiently energetic to disrupt the layers of ions and solvating molecules, or when thermal motion has stirred away the surface accumulation of charges. Typically, if the repulsive barrier is larger than $\sim 10\,kT$ (k = Boltzmann constant), the collisions of particles may not overcome the barrier and agglomeration will not happen.

---

[11] The theory is developed by B. Derjaguin and L. Landau, and independently E. Verwey and J. T. G. Overbeek.

While electrostatic stabilisation is very useful for the preparation of stable colloidal systems, the method is difficult to apply to multiphase systems (e.g. different solids that carry different surface charges), and also to electrolyte-sensitive systems. Furthermore, this process is a kinetic stabilization, and it is almost impossible for the agglomerated particles to be re-dispersed. Alternatively, steric or electrosteric stabilisation using polymers or surfactants are applied in most solution preparations of nanoparticles.

5.3.2   Surfactants and Microemulsions

Surfactant is the acronym for "surface-active agent". These are molecules typically consist of two parts: a long-chain hydrocarbon (hydrophobic tail) and a polar group at one end (hydrophilic head). Some examples of surfactants include salts of carboxylic acid (i.e. soaps, $RCOO^-Na^+$) and alkyl sulfate ($ROSO_2O^-Na^+$), where R = hydrocarbon chain. When these molecules are dissolved in an aqueous system, they will preferentially assemble at the air/aqueous interface, into membrane films, or into micelles such that their hydrophilic heads remain in the aqueous medium while the hydrophobic tails extend into the air or hydrocarbon region (Fig. 5.11).

Surfactants play an important role in forming *microemulsion*, which is a clear and thermodynamically stable dispersion of two immiscible liquids (i.e. oil and water). As mentioned above, the surfactant molecules will assemble into a monolayer film at the oil and water interface. The curvature and rigidity of the film are affected by various parameters such as pressure, temperature, etc.

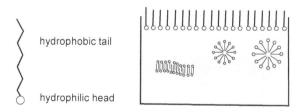

Figure 5.11. Schematics showing one surfactant molecule and the behaviour of surfactant molecules in water forming a monolayer on the surface, or assembling into membranes as well as micelles.

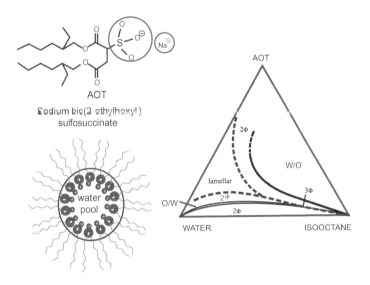

Figure 5.12. An example of a commonly used anion surfactant AOT and a schematic showing the reverse micelles formation. The ternary phase diagram shows the regions of stable phases for a AOT-water-isooctane system.[12]

The relative amounts of the three components can be determined from a ternary phase diagram constructed such as that shown in Fig. 5.12. Depending on which is the dispersed (fine droplets) and the continuous phase, microemulsions are classified as normal micelles (oil droplets in water, o/w) or reverse micelles (water droplets in oil, w/o). Instead of spherical droplets, different structures such as lamellar micelles or vesicles are sometimes formed under suitable conditions.

By optimising the water-to-surfactant molar ratio (also called the *water content*), sizes of the dispersed droplets may be varied within the range of 10-100 nm. Chemists thus make use of the confined space in these droplets for the synthesis of nanoparticles. For example, Pileni and his coworkers have employed an anionic surfactant called AOT (Fig. 5.12) to control the growth of several transition metal nanoparticles in reverse micelles system.[13] In a typical synthesis, precursor salts and reaction initiators/catalysts

---

[12] J. Eastoe, B. H. Robinson, D. C. Steytler and D. Thornleeson, *Advances in Colloid and Interface Science* **36**, 1 (1991).

[13] M. P. Pileni, B. Hickel, C. Ferradini and J. Pucheault, *Chemical Physics Letters* **92**, 308 (1982).

are dissolved separately in aqueous phase. Two microemulsions of the same water content were then prepared and stirred together. The frequent collisions of the reverse micelles would lead to exchange of contents inside the water pool, thus initiating the reaction. The final sizes of the nanoparticles prepared depend mainly on the water content and intermicellar exchange rate. In the literature, many nanoparticles prepared with narrow size distribution have been reported using microemulsion method.

Further Reading

G. Cao, *Nanostructures and Nanomaterials – Synthesis, Properties and Applications* (Imperial College Press, 2004).

P. Atkins and J de Paula, *Physical Chemistry*, 7th Edition (Oxford University Press, 2002).

Exercises

5.1. Once the nanostructures are agglomerated, the smaller the individual nanostructures, the more difficult it is to separate the agglomerates. Explain.

5.2. Molecules can adsorb onto surfaces in two ways — *"Physisorption"* (i.e. physical adsorption) and *"Chemisorption"* (i.e. chemical adsorption). Discuss and distinguish between the two.

5.3. In the discussion of catalytic activity, it is sometimes argued that the activity will be highest when the strength of adsorption on the catalyst surface is intermediate. Try to rationale this argument — why shouldn't the adsorption be strong, and what is the effect of nanometre size in this aspect?

5.4. Discuss briefly the effect of addition of salts into a colloidal mixture.

This page intentionally left blank

Chapter Six

Low-Dimensional Nanostructures

In this chapter, we will move from atoms and molecules to the world of solids. Here, we are less concerned with large scale extended solids, where "large scale" here refers to solids with all dimensions larger than about 100 nm, where size-effects do not play a significant role. Instead, we will concern ourselves with "low dimensional" nanostructures, whereby at least one dimension of the solid is less than about 100 nm in length. In such low dimensional systems and nanostructures, the physical properties differ dramatically from those of their corresponding bulk materials because quantum effects become significant. As seen in Chapter 3, confinement of quantum mechanical wave functions in regions of nanoscale dimensions induces a discretisation of energy levels, and in this chapter we will introduce the effects of low dimensionality on the electron density of states and related electronic properties.

6.1 FROM 3D TO 0D NANOSTRUCTURES

The band theory of solids described in undergraduate solid state physics textbooks is a very successful model for explaining the electronic properties of periodic three-dimensional extended solids. Band theory is based on the assumption that electron properties can be derived by treating the system as a one-electron problem in an average potential determined primarily by the periodic array of ionised atoms in an extended perfect crystal. We will not describe band theory in any detail here, but a useful way

Science at the Nanoscale: An Introductory Textbook
by Chin Wee Shong, Sow Chorng Haur & Andrew T S Wee
Copyright © 2010 by Pan Stanford Publishing Pte Ltd
www.panstanford.com
978-981-4241-03-8

to visualise the difference between conductors, insulators and semiconductors is to plot the available energies for electrons in the materials. Instead of having discrete energies as in the case of free atoms (cf. Chapter 3), the available energy states form bands. An important factor determining electron conduction is whether or not there are electrons in the conduction band.

Figure 6.1 shows a simplified schematic of the energy bands in the three different types of solids — insulators, semiconductors and conductors. In insulators, the electrons in the valence band are separated by a large energy gap from the conduction band, and at normal temperatures no electrons have enough energy to enter the conduction band. In conductors such as metals, the valence band overlaps with the conduction band so substantial numbers of electrons can travel freely in the conduction band, and hence the material conducts electricity. In semiconductors, there is a small gap between the valence and conduction bands, and thermal or other excitations can cause a few electrons to bridge the gap. Since the gap is small, introducing a small amount of a suitable doping material into semiconductors can greatly increase its conductivity, and doping is a key process in the semiconductor industry today.

An important parameter in band theory is the "Fermi level", which is the top of the collection of electron energy levels at 0 K. We shall see later that according to Fermi-Dirac statistics, electrons obey the Pauli exclusion principle, and hence cannot

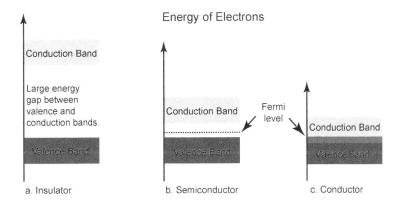

Figure 6.1. A simplified schematic of the energy bands in insulators, semiconductors and conductors.

exist in identical energy states. As such, at 0 K they collect in the lowest available energy states and build up a "Fermi sea" of electron energy states.

The concept of the **Fermi energy** is crucially important in understanding the electrical and thermal properties of solids. Both ordinary electrical and thermal processes involve energies of a small fraction of an electron volt; for instance, at room temperature, the thermal energy kT = 0.026 eV. However, the Fermi energies E_F of metals are of the order of electron volts (see Table 6.1), so most electrons cannot receive energy from such processes since there are no available energy states for them. The **Fermi velocity** is the average velocity of an electron in an atom at 0 K, and is defined by:

$$v_F = \sqrt{\frac{2E_F}{m_e}} \tag{6.1}$$

where m_e is the mass of the electron. The **Fermi temperature** is

Table 6.1 Fermi energy, Fermi temperature, and Fermi velocity of selected metals.

Element	Fermi Energy eV	Fermi Temperature $\times 10^4$ K	Fermi Velocity $\times 10^6$ m/s
Li	4.74	5.51	1.29
Na	3.24	3.77	1.07
K	2.12	2.46	0.86
Rb	1.85	2.15	0.81
Cs	1.59	1.84	0.75
Cu	7.00	8.16	1.57
Ag	5.49	6.38	1.39
Au	5.53	6.42	1.40
Be	14.3	16.6	2.25
Mg	7.08	8.23	1.58
Ca	4.69	5.44	1.28
Sr	3.93	4.57	1.18
Ba	3.64	4.23	1.13
Nb	5.32	6.18	1.37
Fe	11.1	13.0	1.98

defined by:

$$T_F = \frac{E_F}{k} \tag{6.2}$$

where k is the Boltzmann constant. The Fermi level plays an important role in the band theory of solids. In p-type and n-type doped semiconductors, the Fermi level is shifted by the dopant impurities. The Fermi level is referred to as the electron chemical potential in the chemistry context.

6.1.1 Energy Distribution Functions

We next introduce the important statistical mechanics concept of **energy distribution functions**. The distribution function $f(E)$ is the probability that a particle is in energy state E. $f(E)$ is a generalization of the ideas of discrete probability to the case where energy E can be treated as a continuous variable. Three distinctly different distribution functions are found in nature — the Maxwell-Boltzmann, Bose-Einstein, and Fermi-Dirac distribution functions; they are summarised in Fig. 6.2.

The Maxwell-Boltzmann distribution function is a classical function used to describe a system of identical but distinguishable particles, e.g. ideal gas molecules, giving the well-known Maxwell distribution of molecular speeds.

Maxwell-Boltzmann (classical)	$f(E) = \dfrac{1}{Ae^{E/kT}}$	Identical but distinguishable particles, e.g. Molecular speed distribution
Bose-Einstein (quantum)	$f(E) = \dfrac{1}{Ae^{E/kT} - 1}$	Identical indistinguishable particles with integer spin (bosons), e.g. Thermal radiation, specific heat
Fermi-Dirac (quantum)	$f(E) = \dfrac{1}{Ae^{E/kT} + 1}$	Identical indistinguishable particles with half-integer spin (fermions), e.g. Electrons in a metal, conduction in semiconductor

Figure 6.2. The three distinctly different energy distribution functions found in nature — the Maxwell-Boltzmann, Bose-Einstein, and Fermi-Dirac distribution functions. The term A in the denominator of each distribution is a normalization term which may change with temperature.

The Bose-Einstein and Fermi-Dirac distributions differ from the classical Maxwell-Boltzmann distribution because the particles they describe are indistinguishable. Particles are considered to be indistinguishable if their wave packets overlap significantly. This argument arises from the quantum mechanical hypothesis that all particles have characteristic wave properties (cf. de Broglie hypothesis). Two particles can only be considered distinguishable if their separation is large compared to their de Broglie wavelength.

The Bose-Einstein distribution function is used to describe a system of identical and indistinguishable particles with integer spin (bosons), e.g. photons, giving the Planck radiation formula. The Fermi-Dirac distribution function is used to describe a system of identical but indistinguishable particles with half-integer spin (fermions), e.g. electrons. As we shall be focusing mainly on the electronic properties of nanostructures in this chapter, we shall discuss the implications of the Fermi-Dirac function on the electrical conductivity of a semiconductor.

The Fermi-Dirac distribution applies to fermions (e.g. electrons) which must obey the Pauli exclusion principle. Relative to the Fermi energy E_F, it is given by:

$$f(E) = \frac{1}{e^{(E-E_F)/kT} + 1} \tag{6.3}$$

The significance of the Fermi energy is clearly seen at $T = 0$, where the probability $f(E) = 1$ for energies less than the Fermi energy and zero for energies greater than the Fermi energy, i.e. it is a step function. This is consistent with the Pauli exclusion principle which states that each quantum state can have only one particle.

Figure 6.3 shows the Fermi-Dirac function applied to the band structure of a semiconductor. The band theory of solids shows that there is a sizable energy gap between the Fermi level and the conduction band of the semiconductor. At 0 K, no electrons have energies above the Fermi level, and they remain in the valence band since there are no available states in the band gap. At higher temperatures, the Fermi-Dirac distribution is no longer a step function, but has a tail that extends into the conduction band. Hence, some electrons bridge the energy gap and participate in electrical conduction.

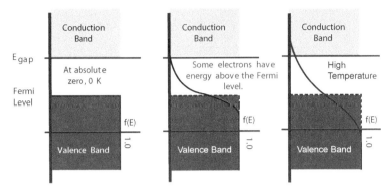

No electrons can be above the valence band at 0 K, since none have energy above the Fermi level and there are no available energy states in the band gap.

At high temperatures, some electrons can reach the conduction band and contribute to electric current.

Figure 6.3. Schematic diagrams of the Fermi-Dirac function applied to the band structure of a semiconductor at different temperatures.

6.1.2 Density of States

Our discussion so far assumes that there is a uniform availability of states for electrons in either the valence or conduction band. The situation is more complicated for real solids and we need to define a **density of states (DOS) function** $g(E)$ to describe the availability of states for electrons to occupy at different energies. The electron population depends upon the product of the Fermi-Dirac function (probability that a given state will be occupied) and the electron density of states. The number of electrons per unit volume with energy between E and $E + \Delta E$ is given by:

$$n(E)\Delta E = g(E)f(E)\Delta E \qquad (6.4)$$

To find out how many ways there are to obtain a particular energy in an incremental energy range dE (the differential limit of ΔE), we use the approach of the quantum mechanical 'particle in a box'. The energy for an infinite walled 3D box (from Eq. 3.30) is:

$$E = \frac{(n_x^2 + n_y^2 + n_z^2)h^2}{8mL^2} \qquad (6.5)$$

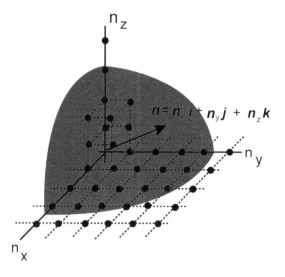

Figure 6.4. Spherical volume of radius R encompassing a number of possible states in positive 3D n-space.

The Rayleigh scheme for counting modes treats the "quantum numbers" n as if they exist in a 3D coordinate space such that a given coordinate of n values (n_x, n_y, n_z) determines a point (hence state) in that space. The number of possible states is proportional to the "volume" in "n-space".

It is convenient to define a radius R in n-space as shown in Fig. 6.4, where:

$$R = \sqrt{n_x^2 + n_y^2 + n_z^2} \tag{6.6}$$

The energy can be expressed in terms of R and vice versa. From Eqs. (6.5) and (6.6):

$$R = \frac{2\sqrt{2mE}L}{h} \tag{6.7}$$

Note that the n-space associated with the particle-in-a-box involves only positive values of n, so the volume must be divided by 8. It must also be multiplied by 2 to account for the two possible spin values of the electron. Hence the total number of available states N is:

$$N = 2\left(\frac{1}{8}\right)\frac{4}{3}\pi R^3 = \left(\frac{8\pi}{3}\right)(2mE)^{3/2}\frac{L^3}{h^3} \tag{6.8}$$

The number of states per unit volume n is:

$$n = \frac{N}{L^3} = \left(\frac{8\pi}{3}\right) \frac{(2mE)^{3/2}}{h^3} \tag{6.9}$$

The final DOS as a function of energy $g(E)$ is the derivative of this population n with respect to energy:

$$g(E) = \frac{dn}{dE} = \frac{4\pi (2m)^{3/2}}{h^3} \sqrt{E} \tag{6.10}$$

This 3D DOS function $g(E)$ represents the number of electron states per unit volume per unit energy at energy E. This expression can be applied to bulk 3D materials, and is independent of the dimension L.

From Eq. (6.4), the carrier density n in a 3D bulk semiconductor can be obtained by integrating the product of the 3D DOS function $g(E)$ and the probability density function $f(E)$ over all possible states, from the bottom of the conduction band E_c, to the top of the conduction band:

$$n = \int_{Ec}^{\infty} n(E)dE = \int_{Ec}^{\infty} g(E)f(E)dE \tag{6.11}$$

The integral in Eq. (6.11) is illustrated by the shaded area in Fig. 6.5. $f(E)$ is a step function with a gentle tail for $T > 0$. $g(E)$ is the 3D DOS function which has the form of a $\sqrt{E}$ function. Hence although the DOS increases with energy (i.e. there are more available states at higher energy), the probability of occupation drops sharply so only the bottom of the conduction band is occupied by electrons. The actual location of the top of the conduction band does not need to be known as the Fermi function goes to zero at higher energies. This is why the upper limit of the integrals in Eq. (6.11) is replaced by infinity.

Substituting the expressions for $g(E)$ and $f(E)$ in Eq. (6.11), we obtain the expression for career density n_0 (where subscript 0 indicates that the system is at thermal equilibrium):

$$n_0 = \int_{Ec}^{\infty} \frac{4\pi(2m_e^*)^{3/2}}{h^3} \sqrt{E - E_c} \frac{1}{1 + e^{\frac{E-E_F}{kT}}} dE \tag{6.12}$$

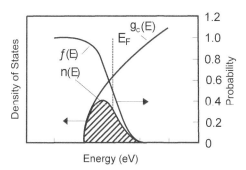

Figure 6.5. Plot showing the functions $f(E)$ and $g_c(E)$; the integral of the product of both functions with respect to energy [Eq. (6.11)] is illustrated by the shaded area.

where m_e^* is the effective electron mass. While this integral cannot be solved analytically at non-zero temperatures, it is possible to obtain either a numeric or an approximate analytical solution.

6.1.3 3D, 2D, 1D, 0D DOS Functions

We have so far derived the 3D DOS function in Eq. (6.10) by counting states in 3D n-space. We can rewrite this expression relative to some reference minimum energy E_{min}, where $E \geq E_{min}$, as:

$$g_{3D}(E) = \frac{dn_{3D}}{dE} = \frac{4\pi (2m^*)^{3/2}}{h^3} \sqrt{E - E_{min}} \qquad (6.13)$$

In low-dimensional nanostructures, we can limit one or more dimensions to nanoscale lengths thereby confining the states in that dimension. Figure 6.6 shows examples of 2D (e.g. self-assembled molecular monolayer), 1D (e.g. molecular wires) and 0D (e.g. quantum dot) nanostructures.

For a 2D nanostructure, we can count states in 2D n-space (a 2D plane) and obtain the corresponding expression for the 2D DOS function. This can be easily shown (left as an exercise at end of chapter) to be:

$$g_{2D}(E) = \frac{dn_{2D}}{dE} = \frac{4\pi m^*}{h^2} \qquad (6.14)$$

Note that the 2D DOS function is a constant independent of energy.

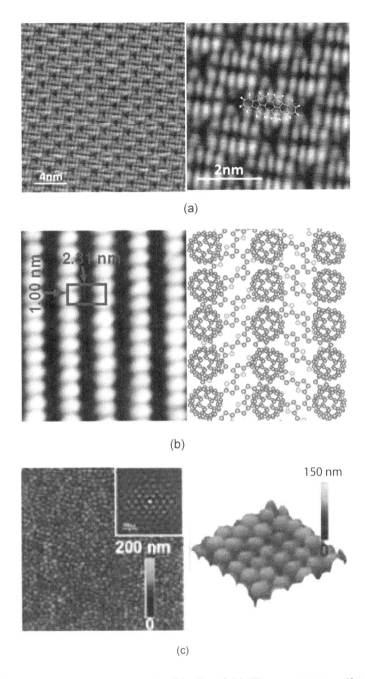

Figure 6.6. Examples of (a) 2D, (b) 1D and (c) 0D nanostructures (from author's lab).

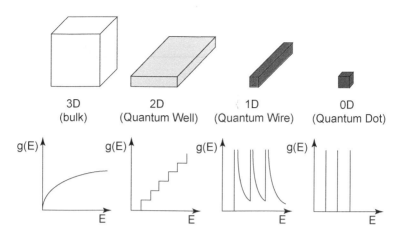

Figure 6.7. Density of states for 3D, 2D, 1D, and 0D structures showing discretization of energy levels and discontinuity in the density of states.

For a 1D nanostructure such as the carbon nanotube, we can also count states in 1D n-space (a 1D line) and obtain this expression for the 1D DOS function:

$$g_{1D}(E) = \frac{dn_{1D}}{dE} = \sqrt{\frac{2m^*}{h^2}} \frac{1}{\sqrt{E - E_{\min}}} \qquad (6.15)$$

In a 0D nanostructure, there is only one discrete energy in 0D space. All the available states therefore exist only at discrete energies and can be represented by a delta function. In real quantum dots, however, the size distribution of the ensemble of quantum dots leads to a broadening of this delta function. Figure 6.7 shows schematics of the 3D, 2D, 1D and 0D DOS functions. Note that there may be different quantised levels in low dimensional nanostructures, e.g. even though the 2D DOS is constant for a quantum well, it is usually a step function with steps occurring at the energy of each quantized level. Figure 6.8 shows the calculated DOS for different types of carbon nanotubes, revealing the characteristic 1D DOS features.

Table 6.2 shows the number of degenerate states for the ten lowest energy levels in a quantum well (2D), quantum wire (1D) and quantum box (0D). In the 2D quantum well, the ratio of the ith energy level to the ground state level E_0 is proportional to n^2, and in each case there is only one way in which n can be arranged to obtain this energy; hence the degeneracy of each energy level is 1.

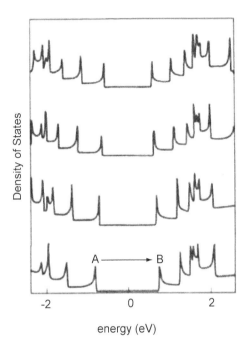

energy (eV)

The one-dimensional density of states for (8,8), (9,9), (10,10), and (11,11) armchair nanotubes show discrete peaks at the positions of the one-dimensional band maxima or minima. For these metallic nanotubes, the density of states is non-zero at E= 0. Optical transitions can occur between mirror-image spikes, such as A ➔ B

Figure 6.8. Calculated 1D DOS of different types of carbon nanotubes. [Reprinted with permission, M. Dresselhaus, G. Dresselhaus, P. Eklund and R. Saito, Carbon nanotubes, *Physics World*, January 1998, pp. 33–38.]

The 1D quantum wire has two values of n to determine its energy; hence the energy is proportional to the sum of the squares of each n value, e.g. $5 = 1^2 + 2^2$; where the values of n are different there are two ways of obtaining the same ratio of energy, i.e. the degeneracy is two. The 0D quantum box is confined in all three n-directions and its energy is proportional to the sum of the squares of the n values. The number of degenerate energy levels is also greater, e.g. when the n values are all different, there are six ways in which the n values can be arranged to produce the same value of energy.

Table 6.2 The table summarizes the ratio of allowed energies to ground state energy and degeneracy of the energy level for 2D, 1D and 0D structures.

State	2D		1D			0D		
	E/E_0	$n(E)$	E/E_0	Degenerate States	$n(E)$	E/E_0	Degenerate States	$n(E)$
1	1	1	2	(1,1)	1	3	(1,1,1)	1
2	4	1	5	(2,1),(1,2)	2	6	(2,1,1),(1,2,1),(1,1,2)	3
3	9	1	8	(2,2)	1	9	(2,2,1),(1,2,2),(2,1,2)	3
4	16	1	10	(3,1),(1,3)	2	11	(3,1,1),(1,3,1),(1,1,3)	3
5	25	1	13	(3,2),(2,3)	2	12	(2,2,2)	1
6	36	1	17	(4,1),(1,4)	2	14	(3,2,1),(3,1,2),(2,3,1) (2,1,3),(1,3,2),(1,2,3)	6
7	49	1	18	(3,3)	1	17	(3,2,2),(2,3,2),(2,2,3)	3
8	64	1	20	(4,2),(2,4)	2	18	(4,1,1),(1,4,1),(1,1,4)	3
9	81	1	25	(4,3),(3,4)	2	21	(4,2,1),(4,1,2),(1,4,2) (1,2,4),(2,4,1),(2,1,4)	6
10	100	1	26	(5,1),(1,5)	2	27	(3,3,3)	1

6.2 ELECTRON TRANSPORT PROPERTIES IN LOW DIMENSIONAL SYSTEMS

For bulk 3D materials, the electric current in a material is proportional to the voltage across it, and the material is said to be "*ohmic*", i.e. it obeys *Ohm's law* ($V = IR$). A microscopic view suggests that this proportionality ($V \propto I$) comes from the fact that an applied electric field superimposes a small drift velocity on the free electrons in a metal. For ordinary currents, this drift velocity is on the order of *mm per second*, which is much slower than the speed of the electrons $\sim$ *a million metres per second*. The electron speeds are themselves small compared to the speed of *transmission of an electrical signal* down a wire, which is of the order of the speed of light, *300 million metres per second*. The current density (electric current per unit area, $J = I/A$) can be expressed in terms of the free electron density as:

$$J = nev_d \tag{6.16}$$

where n is the free electron density, e the electron charge, and v_d the electron drift velocity. From Ohm's law, and expressing resistance in terms of *conductivity* σ or *resistivity* ρ ($R = \rho L/A$),

$$J = \frac{V}{RA} = \frac{V}{\dfrac{\rho L}{A} A} = \frac{E}{\rho} = \sigma E \tag{6.17}$$

which is Ohm's law expressed in terms of current density J and electric field E.

6.2.1 *2D Electron Transport*

Electrons in a large block of material are free to travel in any direction, forming a 3D "electron gas". If we create a thin slab of the material, the electrons can still travel freely in the plane of the slab, but their motion in the third dimension is restricted. The wave function of an electron in this dimension is represented by a standing wave. The situation is analogous to the "particle-in-the-box" concept introduced in Chapter 3, whereby a particle is confined between two rigid walls of infinite potential energy from which it cannot escape. The motion of the electron in the third dimension is quantized and can be represented by a "ladder" of

Table 6.3   Resistivity of common materials at 20°C.

Material	Resistivity ρ (ohm.m)
Silver	1.59×10^{-8}
Copper	1.68×10^{-8}
Aluminum	2.65×10^{-8}
Tungsten	5.6×10^{-8}
Iron	9.71×10^{-8}
Platinum	10.6×10^{-8}
Lead	22×10^{-8}
Mercury	98×10^{-8}
Carbon* (graphite)	$3\text{-}60 \times 10^{-5}$
Germanium*	$1\text{-}500 \times 10^{-3}$
Silicon*	$0.1 - 60$
Glass	$1\text{-}10{,}000 \times 10^{9}$

*The resistivity of semiconductors depends strongly on the presence of impurities in the material, a fact which makes them useful in solid state electronics.

levels of increasing energy, with the separation between the levels growing larger as the slab is made thinner. The electrons can occupy any of the levels that lie below the maximum or Fermi energy E_F.

Two-dimensional electron gases can also be artificially created using a number of semiconductors. A widely used system is the gallium arsenide-aluminium gallium arsenide (GaAs/AlGaAs) heterostructure which can be grown to near-epitaxial perfection using the molecular beam epitaxy (MBE) technique. It contains a 2D electron gas at the interface between the two materials, which have similar properties. There is little disorder in this region, which means that electrons are scattered much less than they are in silicon and are highly mobile.

Figure 6.9 shows a GaAs/AlGaAs/GaAs heterostructure where the conduction bands of GaAs and AlGaAs are offset from each other allowing electrons to collect in GaAs but not in AlGaAs. To provide the electrons, the middle of the AlGaAs region is silicon-doped. These donors become positively ionised and provide electrons which collect in the GaAs just at the interface, since they are attracted to the positive ions. They distort the conduction band as shown, forming a triangular "well" at the interface which

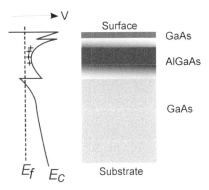

Figure 6.9. (Right) Cross-section through a GaAs/AlGaAs/GaAs heterostructure grown by MBE with nearly atomically sharp interfaces. (Left) The corresponding band diagram, i.e. the energy of the conduction band (the lowest energy electrons can have). The dashed line is the Fermi energy (roughly defined as the highest energy that electrons can have in equilibrium).

goes slightly below the Fermi energy so that electrons can collect there. This well is so narrow that all the electrons there behave as quantum-mechanical waves, with the same wave function in the vertical direction. Thus the only degrees of freedom for the electrons are in the plane of the interface, and so they are effectively a two-dimensional electron gas.

The GaAs/AlGaAs system plays a valuable role in the modern optoelectronics industry and has been used in a number of important experiments in physics, most notably the discovery in 1982 of the "fractional quantum Hall effect" by Daniel Tsui, Horst Stormer and Arthur Gossard at Bell Laboratories, USA.[1] Tsui and Stormer shared the 1998 Nobel Prize for Physics with Robert Laughlin for "their discovery of a new form of quantum fluid with fractionally charged excitations". The quantum Hall effect is a quantum-mechanical version of the Hall effect, observed in two-dimensional electron systems subjected to low temperatures and strong magnetic fields, in which the Hall conductance σ takes on the quantised values:

$$\sigma = \nu \frac{e^2}{h} \tag{6.18}$$

[1] D. C. Tsui, H. L. Stormer and A. C. Gossard, *Phys. Rev. Lett.* **48**, 1559 (1982).

where e is the elementary charge and h is Planck's constant. In the integer quantum Hall effect, ν takes on integer values ($\nu = 1, 2, 3$, etc.). However, in the fractional quantum Hall effect, ν can occur as a vulgar fraction ($\nu = 2/7, 1/3, 2/5, 3/5, 5/2$ etc.).

The quantisation of the Hall conductance has the important property of being extremely precise. Actual measurements of the Hall conductance have been found to be integer or fractional multiples of e^2/h to nearly one part in a billion. This phenomenon, referred to as "exact quantisation", has been shown to be a subtle manifestation of the principle of gauge invariance. It has allowed for the definition of a new practical standard for electrical resistance — the resistance unit h/e^2, or approximately 25812.8 ohms, is referred to as the *von Klitzing constant* R_K, after Klaus von Klitzing, the discoverer of exact quantisation. In 1980, von Klitzing made the unexpected discovery that the Hall conductivity was exactly quantised,[2] and for this finding, von Klitzing was awarded the 1985 Nobel Prize in Physics.

The 2D electron gas in a GaAs-AlGaAs heterojunction has a Fermi wavelength which is a hundred times larger than in a metal. This makes it possible to study a constriction with an opening comparable to the wavelength, and much smaller than the mean free path for impurity scattering. Such a constriction is called a *quantum point contact*. In 1988, the Delft-Philips and Cambridge groups reported the discovery of a sequence of steps in the conductance of a constriction in a 2D electron gas, as its width W was varied by means of a voltage on the gate.[3] The experimental step size is *twice* e^2/h because spin-up and spin-down modes are degenerate.

6.2.2 1D Electron Transport

In a 1D quantum wire, electrons are now quantum mechanically confined in two dimensions, and can only travel freely in one dimension. In 1957 Rolf Landauer showed that the electrical conductance ($G = 1/R$) of a 1D quantum wire where electrons travel

---

[2] K. von Klitzing, G. Dorda and M. Pepper, *Phys. Rev. Lett.* **45**, 494 (1980).

[3] B. J. van Wees *et al.*, *Phys. Rev. Lett.* **60**, 848 (1988); *Phys. Rev. B* **43**, 12431 (1991).

ballistically is given by:

$$G = \frac{2e^2}{h} \tag{6.19}$$

where e is the electron charge and h is Planck's constant. Hence, 1D ballistic electron transport is quantised and the quantum of resistance $R = 1/G = h/2e^2 \approx 12.9k\Omega$. In the absence of collisions, the resistance can only originate from the conductor-contact interface, and hence R is often called the *contact resistance*. Ohm's law implies that the conductance is inversely proportional to the length of the sample, but the conductance of ballistic structures is independent of the length of the sample. Such 1D quantised conductance has been observed in individual multiwall carbon nanotubes (Fig. 6.10).[4]

Electron transport at the nanoscale depends on the relationship between the sample dimensions and three important characteristic lengths:

1. The mean free path L_{fp}, which represents the average distance an electron travels before it collides inelastically with impurities or phonons;
2. The phase relaxation length L_{ph}, which is the distance after which the phase memory of electrons, or electron coherence, is lost due to time-reversal breaking processes such as dynamic scattering;
3. The electron Fermi wavelength λ_F, which is the wavelength of electrons that dominate electrical transport.

The electron transport is diffusive if $L > L_{fp}$. The transport is ballistic if the sample length $L \ll L_{fp}, L_{ph}$, i.e. the electron does not scatter and the electron wave function is coherent. In modern high-mobility semiconductor heterostructures, L_{fp} and L_{ph} can be tens of micrometers; on the other hand for polycrystalline metal films L_{fp} is just tens of nanometres. The conductance G is quantized $G \sim e^2/h$ when $L \sim \lambda_F$. Diffusive transport involves electrons with a wide energy distribution, but ballistic transport involves only electrons close to the Fermi energy, E_F.

The quantisation of an electron's resistance can be understood in semi-classical terms. When a voltage V is applied between

---

[4] S. Frank *et al.* Carbon nanotube quantum resistors, *Science* **280**, 1744–46 (1998).

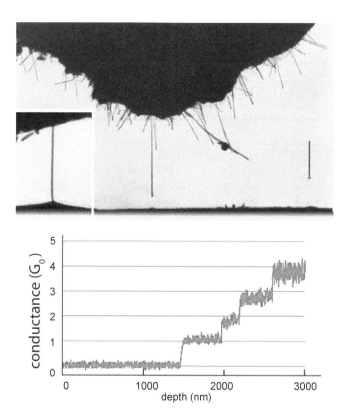

Figure 6.10. Walt de Heer and co-workers measured the conductance of individual multiwall carbon nanotubes. Conductance increases in units of the quantum of conductance as the number of individual nanotubes making contact with the mercury increases, suggesting that nanotubes are ballistic 1D conductors. [Image courtesy of Prof. Walter A. de Heer.]

the source and the drain, it generates a current $I \sim vN(E)$ eV, where v is the velocity of the electrons, $N(E)$ is the density of states and e is the charge on an electron. But since $v \sim \sqrt{E}$ and $N(E) \sim 1/\sqrt{E}$, the two terms cancel and the resistance (V/I) depends only on e and Planck's constant, h. It turns out that each quantised energy level has a "quantised resistance" (or "quantum point-contact resistance") of $h/2e^2 \approx 12.9k\,\Omega$. When we have M levels or modes, each acts independently like resistors in parallel, and the total resistance is simply $h/2Me^2$.

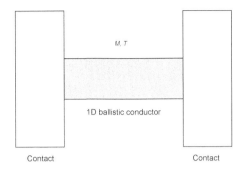

Figure 6.11. Ballistic conductance needs to take into account M parallel subbands and transmission probability T through the conductor.

In a ballistic conductor, there are often a finite number of transverse modes M (or parallel 1D subbands), where M is an integer. This is analogous to a ballistic conductor with variable width, depending on the number of occupied subbands as shown in Fig. 6.11. Furthermore, not all electrons injected at one contact arrive at the other contact, and the electron wave function can be likened to tunneling through a barrier with transmission probability T. Hence, the conductance of a ballistic conductor between two reflectionless contacts at temperature 0 K is given by the Landauer formula:

$$G = \frac{2e^2}{h} MT \tag{6.20}$$

The current between the contacts is therefore:

$$I = GV = \frac{2e^2}{h} MTV \tag{6.21}$$

We can express the total resistance between the contacts as a sum of the contact resistance and the resistance of the conductor with transmission probability T:

$$R = \frac{h}{2e^2 M} \frac{1}{T} = \frac{h}{2e^2 M} \left(1 + \frac{1-T}{T} \right) = \frac{h}{2e^2 M} + \frac{h}{2e^2 M} \left(\frac{1-T}{T} \right) \tag{6.22}$$

The first term is the contact resistance, and the second term is the resistance of the ballistic conductor. Note that for a perfect conductor with $T = 1$, the second term vanishes.

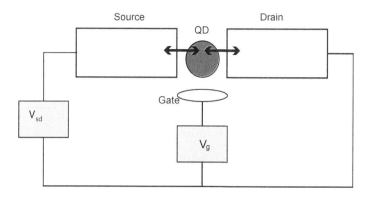

Figure 6.12. Schematic of a quantum dot connected to source and drain contacts by tunnel junctions, and to a gate by a capacitor.

6.2.3 0D Electron Transport

Consider the electronic properties of a quantum dot depicted in Fig. 6.12, which is coupled to three terminals. Electron exchange can occur between two adjacent terminals, as indicated by the arrows. These source and drain terminals connect the small conductor to macroscopic current and voltage meters, and the third terminal provides an electrostatic or capacitive coupling. The number of electrons on this island is an integer N, i.e. the charge on the island is quantised and equal to Ne. If we now allow tunneling to the source and drain electrodes, then the number of electrons N adjusts itself until the energy of the whole circuit is minimised.

When tunneling occurs, the charge on the island suddenly changes by the quantised amount e. An extra charge e changes the electrostatic potential by the charging energy $E_C = e^2/C$, where C is the capacitance of the island. This charging energy becomes important when it exceeds the thermal energy $k_B T$. A second requirement is that the barriers are sufficiently opaque such that the electrons are located either in the source, in the drain, or on the island. This means that quantum fluctuations in the number N due to tunneling through the barriers is much less than one over the time scale of the measurement. This time scale is roughly the electron charge divided by the current. This requirement translates to a lower bound for the tunnel resistances R_t

of the barriers. To see this, consider the typical time to charge or discharge the island $\Delta t = R_t C$. The Heisenberg uncertainty relation: $\Delta E \Delta t = (e^2/C) R_t C > h$ implies that R_t should be much larger than the resistance quantum $h/e^2 = 25.813 \, \text{k}\Omega$ in order for the energy uncertainty to be much smaller than the charging energy.

To summarise, the two conditions for observing effects due to the discrete nature of charge are:

$$R_t \gg \frac{h}{e^2} \tag{6.23}$$

$$\frac{e^2}{C} \gg k_B T \tag{6.24}$$

The first criterion can be met by weakly coupling the dot to the source and drain leads. The second criterion can be met by making the dot small. Recall that the capacitance of an object scales with its radius R. For a sphere, $C = 4\pi \varepsilon_r \varepsilon_0 R$, while for a flat disc, $C = 8\varepsilon_r \varepsilon_0 R$, where ε_r is the dielectric constant of the material surrounding the object.

The circuit with the quantum dot in Fig. 6.12 forms the basis of a *single electron transistor* (SET). The SET has two tunnel junctions sharing one low self-capacitance quantum dot, whose electrical potential can be tuned by the gate, which is capacitively coupled to the dot. The energy levels of the island electrode are evenly spaced with a separation of ΔE. ΔE is the energy needed for each subsequent electron to tunnel to the dot.

The SET has effectively two states:

1. *Blocking state*: As seen in Fig. 6.13(a), no accessible energy levels are within tunneling range of the electron (red) on the source contact. All energy levels on the island electrode with lower energies are occupied.

2. *Positive voltage applied to gate electrode*: Energy levels of the island electrode are lowered and the electron (green 1) can tunnel onto the island (2), occupying a previously vacant energy level. From there it can tunnel onto the drain electrode (3) where it inelastically scatters and reaches the drain electrode Fermi level (4).

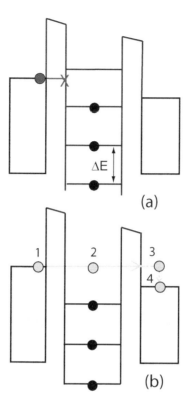

Figure 6.13. Energy level diagram of the single electron transistor.

6.3 QUANTUM DOTS, QUANTUM WIRES

In this final section, we describe some common methods of synthesizing low dimensional nanostructures. In general, the approaches adopted to fabricate nanostructures can be classified into top-down and bottom-up methods. In the top-down approach, the size of bulk materials is gradually reduced to the nanometre scale by using etching techniques with the help of various lithography techniques. This method has proven to be a great success in the highly developed semiconductor industry. On the other hand, the bottom-up method utilises the self-assembly of basic building blocks into nanostructures and such a process can be controlled by the growth dynamics. Depending on the form of the building blocks, the bottom-up method can be further divided into solution-based or vapour-based routes.

6.3.1 *Quantum Dots*

Semiconductor nanoparticles or quantum dots are normally prepared chemically via solution-based routes, often at elevated temperatures and sometimes at elevated pressures (hydro- or solvo-thermal methods). The most commonly studied quantum dots include metal sulfide or metal selenide compounds such as CdS, CdSe, InSe, PbS, ZnS, etc.

In the chemical synthesis of binary semiconductors, two precursors each containing one of the two elements are typically used as the starting reactants. A suitable solvent system is needed for good mixing of the two reactants for a homogeneous reaction to occur. In many cases, the injection method is used whereby a solution of one precursor is injected into a hot solution of the second precursor. In such cases, a burst of nuclei is generated instantaneously in the hot solution when the reactants are mixed together. These nuclei are allowed to grow in the hot solution (ageing) and the reaction is quenched at the appropriate time to give the desired nano-sized particles. The basics of the nucleation and growth processes will be further discussed in Chapter 7.

In order to reduce the tendency to aggregation, organic surface capping agents (as discussed in Chapter 5) are often added in the preparation of quantum dots. Hence, surface-capped quantum dots of varying sizes have been isolated through such colloidal methods and were found to display interesting colours as shown in Fig. 6.14. In the quantum confined regime, the control of particle sizes allows the band gap to be "tuned" to give the desired electronic and optical properties.

The nature and amount of organic surface capping agents used have been found to be crucially important in solution-based routes of nanostructure fabrication. Typically, smaller sized quantum dots are formed when the amount of capping agent added is increased. In some cases, the added capping agent also plays a second role in influencing the kinetics of the chemical processes concerned. For example, when silver(I) thiobenzoate (AgTB) precursor was decomposed to produce Ag_2S nanocrystals, the use of hexadecylamine (HDA) as capping agent was found to also affect the morphology of the produced nanocrystals.[5] A combined

---

[5] W. P. Lim, Z. Zhang, H. Y. Low and W. S. Chin, *Angew. Chem. Int. Edn*, **43**, 5685 (2004).

Figure 6.14. Various sizes of CdSe nanoparticles and their solutions. The bulk solids are typically black in colour. [Image courtesy of Prof. O'Brien, The Manchester Materials Science Centre, UK.]

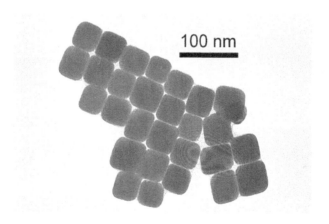

Figure 6.15. Ag_2S cubic nanocrystals produced at $120^\circ C$ and [HDA]/[AgTB] ratio of 8 (from author's lab).

tuning of HDA-to-AgTB molar ratio as well as the reaction temperature induces the formation of interesting Ag_2S nanocubes (Fig. 6.15).

It has also been demonstrated that interesting nanostructures such as CdSe and CdTe tetrapods can be cleverly prepared via the addition of capping agents that specifically bind with particular facets of the growing nanocrystals.[6] Such tunability of the shape and band-gap is very attractive for applications in nanocrystal-based solar cells.

6.3.2 Quantum wires

Surface melting assisted oxidisation can be used to directly grow metal oxide nanostructures without the presence of solution or vapour. Here, we shall detail the commonly used vapour based method and introduce a simple and straightforward thermal heating technique for the growth of 1D nanowires.

The vapour-based route is a widely adopted approach for the fabrication of metal oxide nanostructures. The production of vapour can be achieved in many ways including simple thermal evaporation, laser ablation, sputtering, arc discharge, etc. Typically, the precursor vapour is transported by a gas flow from source to the deposition substrate at a certain temperature range. The process is often implemented in a furnace chamber, such as a tube furnace. A typical setup for vapour-based technique via tube furnace is schematically shown in Fig. 6.16.

In this setup, the precursor material is placed in a tungsten or ceramic boat and positioned at the center of the tube in the furnace. Substrates are placed downstream at lower temperature ranges to collect the oxide products. Depending on different combinations of growth conditions, the final products could be different. In this method, most products are metal oxide

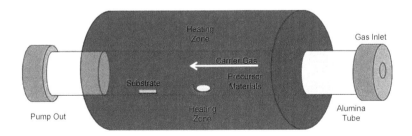

Figure 6.16. Typical setup of a tube furnace for the growth of nanowires using vapour based method.

---

[6] L. Manna, D. J. Milliron, A. Meisel, E. C. Scher and A. P. Alivisatos, *Nature Materials* **2**, 382 (2003).

nanostructures. In the furnace, parameters such as pressure, gas flow rate and temperature gradient are varied to optimise the growth. In addition, the substrates are sometimes decorated with catalytic nanoparticles to promote the growth of 1D nanowires. A wide variety of nanowire products have been achieved by various research groups. Depending on the type of nanowires fabricated, different mechanisms for the growth of the nanostructure have been proposed. These include the vapour-solid (VS) growth and vapour-liquid-solid (VLS) growth mechanisms. Direct vapour-solid (VS) growth is one of the simplest routes to fabricate nanowires. In this technique, the vapour is generated by evaporating source materials or decomposing precursor components, which then condense on target substrates at lower temperatures.

VLS growth has achieved great success in the fabrication of nanostructures with high crystal quality and in large quantities. This method makes use of metallic catalysts, which form eutectic liquids with the source materials at the appropriate temperatures. The precursor vapour dissolves into liquid drops that eventually become saturated. Solid nuclei precipitate after saturation and subsequently continue to grow into nanostructures. Thus VLS growth is often characterised by the presence of catalyst particles located at the tips of nanostructures. VLS growth has been used to synthesise a rich variety of inorganic nanostructures including elemental semiconductors, III-V semiconductors, and II-VI semiconductors. Nanostructures of metal oxides such as ZnO, MgO, TiO_2, SnO_2 etc. have also been synthesised by the VLS method. One of the main advantages of VLS growth is that the diameter and position of 1D nanostructures can be controlled by the size and position of the catalysts. Thus, the 1D nanostructure products can be highly uniform in diameter and readily patterned. In addition, the nanostructures fabricated can be well aligned. Figure 6.17 shows a SEM image of aligned array of ZnO nanowires synthesised on Si substrates.

Nanoscale metal oxide materials with fascinating morphologies can also be simply synthesised by heating pure metallic foils or plates on a hotplate in ambient or appropriate atmospheres. Such a method was described as early as the 1950s by Pfefferkron[7] and

---

[7] G. Pfefferkorn, *Umschau Wiss u. Tech.* **21**, 654 (1954).

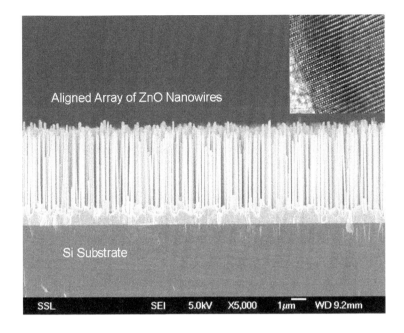

Figure 6.17. Cross sectional SEM image of aligned array of ZnO nanowires grown on Si substrate. The inset shows a HRTEM image of a single ZnO nanowire showing the high crystalline quality of the nanowires. (from author's lab).

Arnold.[8] They described a method where metal oxide filaments were created by annealing corresponding metals in air at elevated temperatures. We shall discuss this cost-effective technique here.

As a typical example, aligned CuO nanowire arrays have been synthesised by heating Cu plates on a hotplate at elevated temperatures. The technique is illustrated in Fig. 6.18. Firstly, polished Cu metal pieces (Figs. 6.18(a), (b)) are placed on the hotplate at a typical temperature of 400°C. Soon after the onset of heating, the shining metallic pieces start to turn dull and darken. The heating is continued for the desired duration and then cooled naturally to room temperature. Figure 6.18(d) shows a cross sectional SEM image of the surface of the heated Cu foil showing an aligned array of CuO nanowires. The height, diameter and density of the nanowire arrays can be controlled by the heating temperature

[8] S. M. Arnold and S. E. Koonce, *J. Appl. Phys.* **27**, 964 (1956).

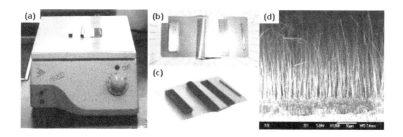

Figure 6.18. Illustration of a simple method to fabricate aligned CuO nanowires showing (a) hotplate with polished Cu plates and tubes; (b) freshly prepared Cu plates and tube; (c) Cu plates and tubes after heating for 10 minutes @ 400°C; and (d) SEM images of the sample surface showing aligned CuO nanowires (from author's lab).

and duration. TEM, HRTEM and Raman studies have shown that these nanowires are single crystalline monoclinic CuO.[9]

A solid-liquid-solid (SLS) mechanism is proposed to explain the growth of CuO nanostructures on the hotplate. Heating of the sample gives rise to surface melting of the metal resulting in a liquid or quasi-liquid medium on the surface. At the same time, the metal is oxidized to a sub-oxide, acting as the precursor for the growth of final products. The sub-oxides on the surface are further oxidized and precipitate from the liquid media, forming different nanostructures depending on the growth kinetics. Surface diffusion continues to supply the materials for the growth of the nanostructures. In this case, the metal foil acts as both the substrate and source material. The synthesis of oxide nanostructures is useful in some electronic applications such as field emission devices, since no additional electrical contact is needed for as-grown oxide/metal samples. This method has been successful in the syntheses of other metal oxides such as Fe_2O_3, Co_3O_4 and ZnO giving rise to a rich variety of different morphologies.[10-12]

---

[9] Y. W. Zhu, T. Yu, F. C. Cheong, X. J. Xu, C. T. Lim, V. B. C. Tan, J. T. L. Thong and C. H. Sow, *Nanotechnology* **16**, 88 (2005).

[10] T. Yu, Y. Zhu, X. Xu, K. S. Yeong, Z. Shen, P. Chen, C. T. Lim, J. T. L. Thong and C. H. Sow, *Small* **2**, 80 (2006).

[11] T. Yu, Y. Zhu, X. Xu, Z. Shen, P. Chen, C. T. Lim, J. T. L. Thong and C. H. Sow, *Adv. Mater.* **17**, 1595 (2005).

[12] Y. Zhu, C.-H. Sow, T. Yu, Q. Zhao, P. Li, Z. Shen, D. Yu and J. T.-L. Thong, *Adv. Funct. Mater.* **16**, 2415 (2006).

Further Reading

C. Kittel, *Introduction to Solid State Physics* (Wiley, 2005), especially Chapter 18 by P. L. McEuenin in 8th edition only.

John H. Davies, *The Physics of Low-dimensional Semiconductors* (Cambridge, 1998).

L. P. Kouwenhoven, C. M. Marcus, P. L. McEuen, S. Tarucha, R. M. Westervelt, and N. S. Wingreen, Electron transport in quantum dots, *Proceedings of the NATO Advanced Study Institute on Mesoscopic Electron Transport*, edited by L. L. Sohn, L. P. Kouwenhoven and G. Schön (Kluwer Series E345, 1997).

K. Berggren and M. Pepper, New directions with fewer dimensions, *Physics World*, October 2002, 37.

M. Dragoman and D. Dragoman, *Nanoelectronics: Principles and Devices* (Artech House, Boston, 2006).

S. Datta, *Electronic Transport in Mesoscopic Systems* (Cambridge University Press, 1995).

R. Saito *et al.*, *Physical Properties of Carbon Nanotubes* (Imperial College, 1998).

M. S. Dresselhaus *et al.*, *Carbon Nanotubes* (Springer, 2001).

S. Reich *et al.*, *Carbon Nanotubes* (Wiley-VCH, 2004).

P. L. McEuen *et al.*, "Single-walled carbon nanotube electronics," *IEEE Transactions on Nanotechnology*, 1, 78 (2002).

Ph. Avouris *et al.*, "Carbon nanotube electronics", *Proceedings of the IEEE*, 91, 1772 (2003).

Exercises

6.1 (i) Calculate the energy relative to the Fermi energy for which the Fermi function equals 5%. Write the answer in units of kT. (ii) For intrinsic (undoped) silicon with a band gap of 1.1eV at 1500 K, what is the population of conduction electrons (m^{-3})? Comment on your result. Note that the melting point of silicon is 1687 K, and atom density of silicon is 5×10^{28} atoms m^{-3}.

6.2 Calculate the number of states per unit energy in a 100 by 100 by 10 nm piece of silicon ($m^* = 1.08m_0$) 100 meV above the conduction band edge. Write the result in units of eV^{-1}.

6.3 Show that density of states functions in 3D, 2D and 1D is given by the expressions:

$$g_{3D} = \frac{dn_{3D}}{dE} = \frac{8\pi\sqrt{2}}{h^3} m^{*3/2} \sqrt{E - E_{min}}, \, E \geq E_{min}$$

$$g_{2D} = \frac{dn_{2D}}{dE} = \frac{4\pi m^*}{h^2}, \, E \geq E_{min}$$

$$g_{1D} = \frac{dn_{1D}}{dE} = \sqrt{\frac{2m^*}{h^2}} \frac{1}{\sqrt{E - E_{min}}}, \, E \geq E_{min}$$

6.4 The conductance of a ballistic conductor between two reflectionless contacts at 0 K is given by:

$$G = \frac{2e^2}{h} MT$$

Define the symbols in the equation, and account for the contributions to the total resistance of this system comprising two contacts and a ballistic conductor with transmission of 50% between the contacts.

6.5 For a room temperature Coulomb blockade device to be constructed using a spherical quantum dot, estimate how small the quantum dot should be.

Hint: For a sphere (QD):

$$C = \frac{Q}{V} = \frac{Q}{\frac{Q}{4\pi\varepsilon_0 R}} = 4\pi\varepsilon_0 R$$

where ε_0 is the permittivity of free space (8.854×10^{-12} F/m).

This page intentionally left blank

Chapter Seven

Formation and Self-Assembly at the Nanoscale

We have briefly discussed the *bottom-up* approach (Fig. 7.1) for the preparation of quantum dots and quantum wires in Section 6.3. In this chapter, let us look at some basic details of this approach. This will require some fundamental concepts in *thermodynamics* — i.e. the study of energy exchanges between physical systems.

7.1 SOME BASIC THERMODYNAMIC DEFINITIONS

7.1.1 Gibbs Energy

In Section 5.1, the Gibbs energy (G) was mentioned briefly when explaining the concept of surface energy. We will now take a closer look at its properties with reference to this fundamental thermodynamic relationship:

$$\Delta G = \Delta H - T\Delta S \qquad (7.1)$$

The notation Δ denotes a change; thus ΔG is the change in Gibbs energy, ΔH is the change in *enthalpy* and ΔS is the change in *entropy*.

Equation 7.1 is applicable for a process occurring at temperature T and constant pressure, e.g. the preparation of quantum dots in an open vessel. Under constant pressure condition, ΔH is equivalent to the heat supplied to the system. When heat is absorbed by a process, ΔH is positive and the reaction is known as an *endothermic*

Science at the Nanoscale: An Introductory Textbook
by Chin Wee Shong, Sow Chorng Haur & Andrew T S Wee
Copyright © 2010 by Pan Stanford Publishing Pte Ltd
www.panstanford.com
978-981-4241-03-8

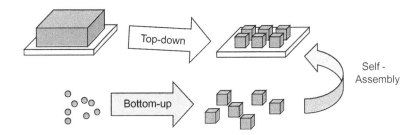

Figure 7.1. Schematic showing the top-down and bottom-up approaches coupled with self-assembly in nanomaterials fabrication.

process. Conversely, when heat is released and ΔH is negative, it is an *exothermic* process.

Entropy may be simply taken as a measure of the randomness of objects in a system. An increase in entropy thus indicates the system has become more *disordered*. Hence ΔS is positive when ice melts to liquid water or when water is vaporized to steam, since the molecules have more freedom to move about in the final state after both processes. The concept of ΔS will be further illustrated when we discuss self-assembly in Section 7.3.

All the three parameters G, H and S are *state functions*, which mean that ΔG, ΔH and ΔS are only defined by the initial and final states of the system, regardless of the thermodynamic path taken. This property allows the values of ΔG, ΔH and ΔS for new processes to be determined from several known processes as long as the initial and final states of the substances are the same, i.e. A → B can be worked out from A → X → Y → Z → B.

The Gibbs energy is often used to assess the direction of a natural process. At constant pressure and temperature, a chemical transformation or process will occur spontaneously in the direction of decreasing G (i.e. $\Delta G < 0$). Thus, it follows from Eq. 7.1 that a spontaneous endothermic reaction, e.g. the boiling of water at 100°C, occurs only if the increase in ΔS outweighs the increase in ΔH. When $\Delta G = 0$, the system is in a state of *equilibrium*, with the forward and backward processes occurring to the same extent.

7.1.2 Chemical Potential

Besides its dependence on temperature (T) and pressure (P), the Gibbs energy also depends on the amount of substance present in the system. The molar[1] Gibbs energy is known as the *chemical potential* (μ) and defined as the partial derivative of G with respect to the number of particles n at constant T and P:

$$\mu = \left(\frac{\partial G}{\partial n}\right)_{T,P} \tag{7.2}$$

For a system of binary mixture with substances X and Y at constant T and P, each substance has its own chemical potential (i.e. μ_X and μ_y) and we can express the change in Gibbs energy in terms of the compositional changes of Δn_x and Δn_y respectively:

$$\Delta G = \mu_X \Delta n_X + \mu_Y \Delta n_Y \tag{7.3}$$

It follows that if a one-component system is at equilibrium between two phases, e.g. liquid X and its vapour, the chemical potentials of the two phases must be equal:

$$\mu_X\,(liquid) = \mu_x\,(gas) \tag{7.4}$$

The chemical potential may be viewed as a measure of the driving force that a substance has for bringing about a change in the system. Transformation occurs spontaneously from a region of high μ to a region of low μ, until μ is uniform throughout the system.

7.1.3 Equilibrium in Solution

If a solid A is placed in a solvent B, it will dissolve until the solvent has become *saturated* with the solute A. Equilibrium is now established between the solid A and the solvated A, i.e. the chemical potential of the solid is equal to the chemical potential of the solute in the saturated solution (Fig. 7.2). If we now define $\mu_A{}^*(l)$ as the chemical potential of the pure liquid A, the chemical potential of the solute $\mu_A(l)$ may be expressed in term of the *activity* A in the solution (a_A):

$$\mu_A\,(l) = \mu_A^*\,(l) + RT \ln a_A\,(l) \tag{7.5}$$

Activity is a dimensionless quantity and may be viewed as the "effective concentration" of the species in solution. Activity of

---

[1] One molar quantity of a substance consists of 6.02214×10^{23} (Avogadro's number) constituent units of that substance; e.g. 1 mole of N_2 molecules consists of 6.02214×10^{23} N_2 molecules.

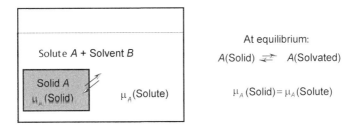

Figure 7.2. Schematic of liquid solute/solid equilibrium.

a pure substance is normally taken as unity. The activity of the solvent is also often assumed to be unity in dilute solution.

Equation 7.5 thus provides a relationship between the Gibbs energy and the concentration for a species in solution. Alternatively, we can also express this relationship in terms of the equilibrium constant K_{eq}:

$$K_{eq} = \frac{a_A (l)}{a_A (s)} = a_A (l) \tag{7.6}$$

$$\Delta G = \Delta G^o + RT \ln K_{eq} \tag{7.7}$$

ΔG^o is defined as the Gibb energy at standard states ($T = 298.15$ K; $P = 101{,}325$ Pa) for concentration at 1 mol dm^{-3}.

7.1.4 Gibbs Energy at the Nanoscale

In Chapter 5, we have learnt that particles at the nanoscale possess a high surface energy (γ) due to their large surface-to-volume ratios. This substantial amount of surface energy is expected to contribute significantly to the Gibbs energy of the nanoparticles. The effect is most evidently observed in the reduction of melting temperatures, which has been found to reduce with the decreasing radius of the nanoparticles.[2] While a quantitative derivation of the contribution of γ to ΔG remains a topic under extensive research, we can at least provide some qualitative considerations as described below.

Let us begin by examining the effect of size on the shift of a chemical equilibrium between reagent particles A_j to product particles B_k. The equilibrium constant at constant P and T

---

[2] P. Buffat, J.-P. Borel, *Phys. Rev. A*, **13**, (1976) 2287.

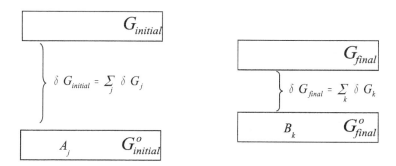

Figure 7.3. Schematic illustration of the surface energy contribution to Gibbs energy for a hypothetical reaction.

provides a measurement for ΔG^o at the standard state:

$$\Delta G^o = -RT \ln K_{eq} \tag{7.8}$$

$$\Delta G^o = G^o_{final} - G^o_{initial} = \sum_k G^o_k - \sum_j G^o_j \tag{7.9}$$

The values of ΔG^o can also be obtained from Eq. 7.1 using standard values of ΔH^o and ΔS^o from any reference handbook, noting that these values are given for the bulk compact substances.

When either or both the reagent and product particles are dispersed into nano-sized entities, there will be an additional contribution of surface energy to Gibbs energy, which we will denote as δG_j and δG_k respectively. Qualitatively, we know that these are dependent on the individual surface area (F) and surface energy (γ) of the particles, e.g.:

$$\delta G_j \propto \gamma_j \times F_j \tag{7.10}$$

The overall Gibbs energy of the reaction can now be rewritten to include the surface contribution:

$$\Delta G = \Delta G^o + \sum_k \delta G_k - \sum_j \delta G_j \tag{7.11}$$

We can see here that there will be a driving force to shift the equilibrium depending on the relative values of $\Sigma \delta G_j$ and $\Sigma \delta G_k$ of the system. As illustrated in Fig. 7.3, if the reagent particles are made very fine and δG_j is larger than δG_k, the reaction occurs readily as now $\Delta G < 0$.

7.2 THE BOTTOM-UP APPROACH

As mentioned in Section 6.3, solution-based routes are commonly employed in the *bottom-up* preparation of quantum dots and nanoparticles. During the preparation, a *supersaturation* of the product species is often generated in the solution. This can be achieved either by lowering the temperature of the saturated solution at equilibrium, and/or by generating a large amount of less soluble product species *in situ*. In this section, we will discuss the homogeneous nucleation and growth processes for the preparation of nanoparticles. Although our discussion will focus on these processes in solution, the fundamental concepts should be equally applicable for growth in gas and solid.

7.2.1 Homogeneous Nucleation

Under the supersaturation condition, the concentration of the solute (C) exceeds its equilibrium solubility (C_o) and the system thus possesses a higher chemical potential according to Eq. 7.5. The system will move toward decreasing G, thus accounting for the driving force for the nucleation and growth processes:

$$\Delta G = -RT \ln \left(\frac{C}{C_o} \right) \tag{7.12}$$

Thus, when $C = C_o$, $\Delta G = 0$ and the system is at equilibrium. When $C > C_o$, ΔG is negative and nucleation should occur spontaneously.

However, the increase in surface energy needs to be counterbalanced during crystal growth. Assuming the nucleus is spherical in shape, there is an increase in chemical potential due to the surface energy (γ) given by:

$$\Delta \mu_s = 4\pi r^2 \gamma \tag{7.13}$$

Since Eq. 7.13 is for a single nucleus of size r, we will rewrite the change in Gibbs energy (Eq. 7.12) in terms of per unit volume (ΔG_v), and express the reduction in chemical potential for the formation of the new spherical nucleus as:

$$\Delta \mu_v = \frac{4}{3} \pi r^3 \Delta G_v \tag{7.14}$$

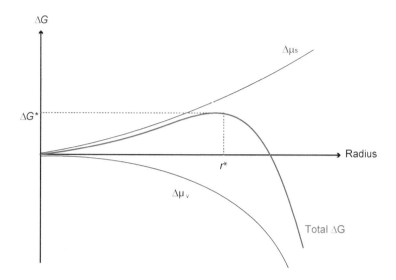

Figure 7.4. Schematic showing the variation of Gibbs energy during the nucleation process.

The overall change in Gibbs energy is thus given by combining equations 7.13 and 7.14:

$$\Delta G = \frac{4}{3}\pi r^3 \Delta G_v + 4\pi r^2 \gamma \tag{7.15}$$

When these equations are plotted as shown in Figure 7.4, we can clearly see that Eq. 7.15 predicts an energy barrier ΔG^* at a critical size r^*. This means that the newly formed nucleus is stable only when $r > r^*$. Below this value, the nucleus will have a natural tendency to re-dissolve into the solution. The value for ΔG^* can be obtained by setting $d(\Delta G)/dr = 0$ when $r = r^*$:

$$\Delta G^* = \frac{16\pi\gamma^3}{3\Delta G_v^2} \tag{7.16}$$

$$r^* = -2\frac{\gamma}{\Delta G_v} \tag{7.17}$$

The nucleation process is often explained by a plot of concentration variation with time as shown in Fig. 7.5. During a typical synthesis, the concentration of the product solute increases as the reaction proceeds. Nucleation of the solute sets in only when ΔG^* is overcome, i.e. when the concentration has reached a critical

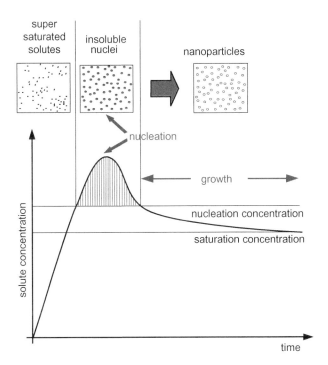

Figure 7.5. Schematic of the nucleation and growth processes with time.[3]

value above saturation. At this point, stable nuclei with sizes above the critical size r^* are formed. After this initial nucleation, the concentration of the species decreases as ΔG decreases further. When the concentration drops below the nucleation point, new nuclei will not form but the fresh solute deposits onto the surfaces of the existing nuclei. The particles then continue to grow until the equilibrium concentration is attained.

The nucleation and growth processes are not separable when the concentration is above the nucleation point. Once nuclei are formed, growth of the nuclei occurs simultaneously as new nuclei are being generated. For the purpose of producing particles with a narrow size distribution, it is advantageous to have all nuclei form within a very short period of time. This is achieved in the injection method (Section 6.3), whereby a short burst of nuclei is formed when the reactants or precursors are introduced into the

---

[3] Adapted from M. Haruta and B. Delmon, *J. Chem. Phys.* **83**, 859 (1986).

solution in one injection. The concentration of reactants decreases quickly immediately after the injection, and the concentration of the solute drops below the nucleation point. The formation of new nuclei is thus prohibited, while all the nuclei formed initially will continue to grow.

7.2.2 The Control of Growth

The eventual size of the nanoparticles produced is dependent on the subsequent growth process. We have discussed in Chapter 5 that smaller particles possess relatively high surface energy; hence they tend to grow larger through processes such as agglomeration, sintering and Ostwald ripening. It is thus important to control or inhibit these processes in order to prepare nano-sized particles.

Ostwald ripening results in the elimination of smaller particles at the expense of larger particles. This phenomenon has been used to produce particles of narrow size distributions in the "*size focusing*" injection method.[4] Basically, at any given solute concentration, there exists a critical size r^* at which the dissolution rate equals to the growth rate. Particles smaller than r^* will dissolve, while particles larger than r^* will grow. When the solute concentration in the solution is depleted due to growth, r^* shifts to a larger value. By injecting additional reagents at the growth temperature, r^* shifts back to a smaller value. Since the diffusion-controlled growth rate is inversely dependent on r, smaller particles grow faster than larger ones. Focusing of the size distribution thus occurs by optimizing injections at suitable intervals.

To effectively prevent agglomeration during the production of nanoparticles, some form of surface stabilization must be devised. Suitable polymers, for example, have been used in the formation of uniformly sized metallic nanoparticles. One example is the preparation of rhodium nanoparticles using polyvinyl alcohol (PVA).[5] The presence of PVA is believed to serve two roles: First, the long chains of polymers form a diffusion barrier to hinder the diffusion of solute species from the surrounding solution; second, these polymers adhere onto the surfaces of the nanoparticles to provide further surface stabilisation.

---

[4] X. Peng, J. Wickham and A. P. Alivisatos, *J. Am. Chem. Soc.* **120**, 5343 (1998).
[5] H. Hirai, Y. Nakao, N. Toshima and K. Adachi, *Chem. Lett.* **905**, (1976).

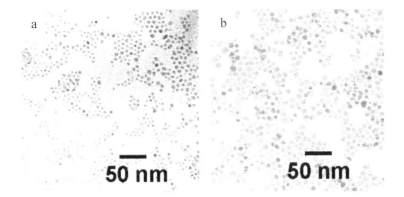

Figure 7.6. CdS nanocrystals produced with (a) higher and (b) lower amounts of the capping molecule hexadecylamine (from author's lab).

Surface adsorption or surface capping agents are used in the synthesis of many semiconductor nanoparticles such as metal sulfides or metal oxides. In these cases, suitable organic molecules are chosen such that they *cap* the surface of the nanoparticles through covalent or dative bonds with the surface atoms of the particles. Examples of such capping agents are: alkanethiols (RSH), carboxylic acids (RCOOH), amines (RNH_x), siloxanes ($RSiOH_3$), etc. (R = alkyl chains). Functional groups such as $-SH$ is known to form strong $-S-Metal$ bonds with transition metals, and the lone pair of nitrogen $-N$: forms dative bonds. Once the functional group chemisorbs onto the particle surface, the alkyl chain R provides steric stabilisation to prevent the approach of other particles and agglomeration. Typically, the final sizes of the nanoparticles prepared are also determined by the amount of capping agent used during the synthesis (see, e.g. Fig. 7.6).

7.3 THE SELF-ASSEMBLY PROCESSES

We have discussed the concept of molecular self-assembly in relation to the formation of supramolecules in Section 4.4. In this section, we will examine the various self-assembly systems and some fundamental aspects of the process.

7.3.1 Thermodynamic Considerations

Self-assembly is a process whereby the basic components in an ensemble (e.g. segments of a molecule, molecules, or small particles) arrange themselves in an orderly manner spontaneously. Nature has created the sophisticated self-assembly structures we observe around us. We have learnt in Section 4.2 that protein molecules spontaneously fold into an aggregate with well-defined 3D shapes. The structure of the stretched-out polypeptide molecule is so precisely sequenced that segments of the molecule interact and fit snugly into each other. The association of protein molecules involves many weak and reversible interactions including hydrogen bonding, van der Waals attraction and hydrophobic interactions.

The interplay between ΔH and ΔS is important in the self-assembly process. As the formation of self-assembly structures brings about a change from disordered components to ordered structures, we expect a decrease in entropy (i.e. negative ΔS). For the process to have a negative ΔG according to Eq. 7.1, the decrease in ΔS must be accompanied by a larger exothermic ΔH. Since the interactions involved in self-assembled systems are relatively weak, the values of ΔH are generally <20 kcal mol^{-1} (cf. Section 4.2). We need to consider two types of entropy loss — *translational* and *conformational* ΔS. The former concerns the loss of freedom in the translational motion of the components when they come together. A rule-of-thumb estimation for the loss in translational entropy for the assembly of components at millimolar concentrations gives $T\Delta S \sim +5.5$ kcal mol^{-1}.[6] When the molecules or components are packed together into an aggregate, freely rotating bonds or conformations may be frozen and this constitutes the conformational entropy loss. The value of this loss depends strongly on the system concerned and the number of mobile segments involved. Overall, since the number of components associating is large, the sum of these unfavorable entropy terms can be significant.

In biological self-assembled systems, the structures of the components are designed to have multipoint contacts and interactions. While the interactions are weak, they work cooperatively to provide a favorable ΔH for the process. On the other hand,

---

[6] G. M. Whitesides, J. P. Mathias and C. T. Seto, *Science*, **254**, 1312 (1991).

the individual components in these systems are able to move with respect to one another. They are "soft" in a sense that the components may adjust their positions within the aggregate, rather than being firmly stuck together or locked into place. Thermodynamically, this moderates the extent of reduction in the entropy term.

Hence, self-assembled biological structures are steady-state structures at the dynamic balance point between order and disorder. For example, folded proteins are stabilised from the unfolded state by only a small energy difference. These self-assembled systems are thus sensitive towards perturbations exerted by the external environment. A small fluctuation that alters the thermodynamic variables, e.g. pH change, may lead to a marked change in or even compromise the structure. The weak cooperative interactions are responsible for the flexibility of the structure and allow for structural rearrangements in the direction determined by ΔG. If fluctuations could bring the thermodynamic variables back to the starting conditions, the structure is likely to revert to its initial configuration. These structures are thus robust and even self-healing.

Since self-assembled structures are made through minimising the Gibbs energy, they are always thermodynamically more stable than the unassembled components. As a consequence, self-assembled structures are also relatively free of defects. For example, components which are not complementary or contain defects are naturally excluded through the self-assembly process in biological systems.

7.3.2 *Self-assembly of Molecules on Surfaces*

Our understanding of how Nature forms self-assembly structures such as proteins and nuclei acids has inspired us to design new and unique supramolecules (Section 4.4). The cell membrane is a bimolecular layer of self-assembled amphiphilic molecules,[7] and has prompted our study of surfactants and micelles (Section 5.3). The potential of self-assembly, however, extends beyond making supramolecular complexes and micelles. Self-assembly is also used as a strategy to produce ordered structures on surfaces.

---

[7] Molecules that have both polar and non-polar segments, e.g. lipids, detergents, surfactants (see Section 5.3).

This topic has attracted much attention as it allows us to engineer special surface architectures required in new technological applications.

Well-defined surfaces, particularly crystalline planes of metallic solids, are found to provide versatile platforms for the assembly of molecules into clusters, chains, 2D arrays, and even 3D superlattice architectures.[8] This is often carried out by *chemical vapour deposition* or *molecular beam epitaxy* inside a high vacuum chamber. The visualisation of these structures is often aided with scanning tunneling microscopy (STM), a nano-tool discussed in Chapter 8. In general, the self-assembly is driven by interactions between the assembled molecules and the substrate surface, as well as between the molecules in adjacent layers. This is the driving force towards the reduction of the overall Gibbs energy.

Researchers have tried to construct complex surface architectures using non-covalent interactions such as hydrogen bonding, $\pi - \pi$ stacking, van der Waals interaction, etc. between neighbouring molecules. In some cases, the first few layers of adsorbed molecules define the architecture capable of *trapping* other entities in the subsequent layer. An example is illustrated in Fig. 7.7 whereby monolayers and bilayers of α-sexithiophene (6T) adsorb on the Ag(111) surface to form stripe-like patterns (Fig. 7.7(a)). In subsequent adsorption experiments, preferential adsorption of C_{60} molecules in linear molecular chains is observed on the bilayer 6T nanostripes (Fig. 7.7(b)).It is proposed that this arises from the donor-acceptor interaction between 6T and C_{60}.

The adsorption of some molecules can also be performed on specific surfaces in solutions to form self-assembled monolayers (SAMs). These are monolayers of amphiphilic molecules that remain intact after the substrates are removed from the solution. These SAMs are stable in air and ordinary temperatures, offering a convenient route to tailor the properties of an entire surface. SAMs can be prepared using different sets of molecules and substrates, examples include alkyl silanes such as octadecyltrichlorosilane (OTS, $CH_3(CH_2)_{17}SiCl_3$) on various oxide surfaces; alkyl carboxylates such as fatty acid on aluminium or mica

[8] J. V. Barth, *Annu. Rev. Phys. Chem.* **58**, 375–407 (2007).

[9] H. L. Zhang, W. Chen, L. Chen, H. Huang, X. S. Wang, J. Yuhara and A. T. S. Wee, "C_{60} molecular chains on α-sexithiophene nanostripes", *Small* **3**, 2015–2018 (2007).

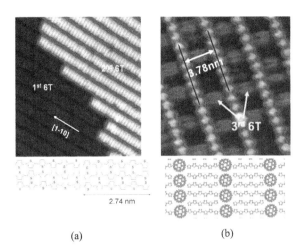

Figure 7.7. Scanning transmission microscopic (STM) images showing adsorption of α-sexithiophene (6T) molecules on Ag(111) surface: (a) self-assembly of mono- and bi-layer of 6T to form nanostripes (30 nm × 30 nm image), (b) self-assembly of C_{60} onto the 6T bilayer patterns (15 nm × 15 nm image) (from author's lab).[9]

surfaces; alkyl thiols on metal surfaces such as gold, silver and copper.

Let us now focus our discussion on one of the most studied SAMs, i.e. the adsorption of alkanethiols on gold surfaces. The simplicity of preparation is illustrated schematically in Fig. 7.8 whereby a clean Au substrate is immersed into a solution of $\sim$0.1–10 mmol L^{-1} alkanethiol in ethanol. The initial adsorption happens in seconds, but the adsorption is left to equilibrate for hours to allow re-organisation into a regular monolayer film. Alkanethiols are widely used because of their good solubility and compatibility with many organic functional groups. The −SH thiol head group *chemisorbs* onto the Au surface through the formation of a strong −S–Au covalent bond. This chemisorption provides the exothermic energy of $\sim$40–45 kcal mol^{-1} needed for the spontaneous process. Alkyl chains of more than eight carbon units are generally found to give stable SAMs. The long hydrocarbon chains maximise inter-chain van der Waals interactions between adjacent molecules.

There have been many studies performed to determine the process by which alkanethiols assemble on Au.[10] The adsorption sites

---

[10] D. K. Schwartz, *Annu. Rev. Phys. Chem.*, **52**, 107 (2001).

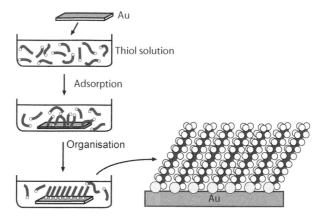

Figure 7.8. Schematic illustration shows the preparation of SAMs using Au substrate and thiol solution.

for the $-$S head groups are at the hollow depressions between three close-packed Au atoms, and the hydrocarbon chains are typically tilted $\sim$30 degrees from the surface normal. It is generally believed that alkanethiol molecules first bind quickly to the Au surface in a 'lying down' manner. When the adsorbed molecules are dense enough on the surface, the alkyl chains interact with each other. They finally adopt an energetically more favorable position in which the alkyl chains are aligned in parallel to each other with their chains fully extended. Over time, islands of adsorbed molecules merge and a full coverage of the SAM is obtained. Sometimes, these SAMs exhibit grain boundaries and defects even after long periods of assembly.

The simple concept of SAM can be applied to create various surface architectures. Critical parameters are the thickness of the monolayer and the composition of the adsorbed films. To build up a second layer, adsorbed molecules carrying specific end groups, e.g. halide groups instead of CH_3 groups, are used so that subsequent layers of molecules can bind directly onto these functional groups in the first monolayer. We can also tailor the interfacial energies of SAMs by changing the structures of the organic molecules.

Engineering surfaces via molecular adsorption or SAMs offers great promise for applications in several different areas. Examples include molecular recognition, selective binding of enzymes to surfaces, corrosion protection, molecular crystal growth, biosensing devices, etc. The self-assembly process is one of the most general strategies available for the formation of regular nanostructures.

7.3.3 Self-assembly of Nano-sized Components

In principle, components of any size can self-assemble just like molecules when the thermodynamic conditions are met. A common example is the precious opal, which is basically a self-assembly of quartz particles ~150 to 300 nm in diameter. These spherical particles adopt a close-packed hexagonal order in crystalline phase. The size distributions and packing order of these particles determine the colour and quality of the precious opal. Light waves scattered from the planes of particles produce interference patterns, thus giving opal its beautiful and iridescent appearance.

Researchers have long demonstrated that regularly sized silica particles can form well-ordered self-assembled structures. Scanning Electron Microscopy (SEM) images show the hexagonal packing order observed for ~800 nm silicate spheres (Fig. 7.9(a)) and a reasonably long-range order obtained in large area (Fig. 7.9(b)). This two-dimensional ordered arrangement is often prepared by spin-casting or solvent evaporation from a solution containing the micron-sized silicate spheres.

Capillary interaction is found to be the main driving force for self-assembly of micron- and nano-sized particles. In the presence of floating or submerged particles, the originally flat liquid surface

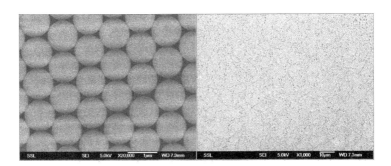

Figure 7.9. Scanning Electron Microscopic (SEM) images showing close-packed assembly of micron-sized silicate nanoparticles. (from author's lab).

deforms as shown schematically in Fig. 7.10. Capillary forces minimize the Gibbs energies by reducing the interfacial areas. The amount of capillary interaction between the two particles is directly proportional to the interfacial deformation created.[11] The extent of deformation is related to the particle size, the surface energy between air and liquid, the properties of the liquid, etc. In the immersion situation, deformation strongly depends on the wetting properties of the two particles and hence the interactions may be adjusted by adding surfactants to the dispersion. An example of the self-assembled arrays of nanoparticles is shown in Figure 7.11. Regular inter-particle distances are observed as

Figure 7.10. Schematics of capillary actions between particles, resulting in their self-assembly.

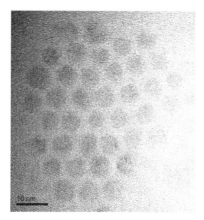

Figure 7.11. Transmission electron microscopy (TEM) images of $\sim$5 nm CdS nanocrystals (prepared with hexadecylamine) self-assemble into a hexagonally closed-packed array (from author's lab).

---

[11] Figure adapted from P. A. Kralchevsky, K. D. Danov and N. D. Denkov, *Handbook of Surface and Colloid Chemistry*, CRC Press, 1997.

these particles are enclosed by surface capping agents with long hydrocarbon chains.

Further Reading

G. Cao, *Nanostructures and Nanomaterials — Synthesis, Properties and Applications* (Imperial College Press, 2004).

C. P. Poole Jr., F. J. Owens, *Introduction to Nanotechnology* (Wiley, 2003).

R. A. L. Jones, *Soft Machines — Nanotechnology and Life* (Oxford University Press, 2004).

Exercises

7.1. The critical radius r^* in Fig. 7.3 presents a practical limit for the smallest size particles that can be prepared in the system. Discuss what other ways that one can use to reduce the critical size as well as the critical Gibbs energy ΔG^*.

7.2. Ellipsometric measurements yield a thickness of 21.1 Å for the adsorbed SAM of hexadecanethiol molecules. An extended hexadecanethiol molecule is estimated to be 24.5 Å in length. What is the tilt angle of the adsorbed monolayer on the substrate?

7.3. A given surfactant molecule has head group with a cross-sectional area of 6 Å^2. A solution of the surfactant in a nonpolar solvent is dispersed slowly onto the surface of water. The volatile solvent evaporates immediately, leaving behind a self-assembly of the surfactant molecules on the surface. (i) Calculate the number of surfactant molecules that will be dispersed in a self-assembled monolayer film of dimensions 15×30 cm. (ii) What is the volume of 0.1 M of the surfactant solution needed to produce the film in (i)?

Chapter Eight

Nanotools and Nanofabrication

One of the fascinating aspects of nanoscience and nanotechnology is that the nanoworld is full of surprising structures and features that one would not have imagined. There have been a number of significant developments that have enabled scientists to explore the world with nanoscale objects. These developments include the development of various types of microscopes, techniques for nanomanipulation, and methods of nanofabrication. This chapter is devoted to discussions of the various types of microscopes, followed by optical tweezers and techniques of nanofabrication. The chapter starts off with a discussion on traditional optical microscope, followed by the scanning electron microscope and transmission electron microscope. Next we focus on the development and working principles of the scanning tunneling microscope and atomic force microscope. The optical tweezer technique, a technique commonly used for the trapping and manipulation of micro/nanoparticles in aqueous suspension, is then introduced. Finally, we present a method for nanofabrication by means of focused laser beams.

8.1 OPTICAL MICROSCOPY

Almost everyone has had the experience of playing with a magnifying glass as a child. Long ago, people found out that the appearance of an object viewed through a piece of transparent crystal becomes larger if the piece of crystal is thicker in the middle. It was also discovered that objects near the end of a tube with

Science at the Nanoscale: An Introductory Textbook
by Chin Wee Shong, Sow Chorng Haur & Andrew T S Wee
Copyright © 2010 by Pan Stanford Publishing Pte Ltd
www.panstanford.com
978-981-4241-03-8

several lenses (compound lenses) appeared much larger than any simple magnifying glass could achieve, and this led to the birth of the optical microscope. Dutchmen Hans and Zacharias Janssen, and Antony van Leuwenhoek were among the pioneers in the development of optical microscopes. Robert Hooke employed his version of the compound microscope to study many types of organisms.

After the early versions of optical microscopes, dramatic progress was achieved in optics and in the construction of the optical microscope. Modern optical microscopes can now achieve very high magnification and are capable of a wide variety of functions. An example of a commercial optical microscope is shown in Fig. 8.1. Optical microscopes are widely used in many areas of science, medicine and engineering, and are the basic indispensible tools in most research laboratories.

With advances in microscopy, the limits of resolution have been steadily improved. (*The resolution of a microscope refers to the separation between two features of an object that can just be distinguished.*) However, optical microscopes cannot achieve ultra high magnifications that are needed to see nanoscale objects such as nanoparticles, molecules and atoms. As most of the optical microscopes

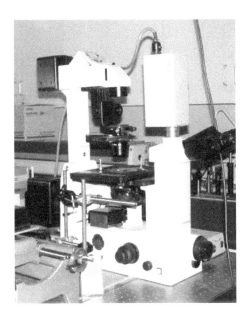

Figure 8.1. A modern optical microscope.

are designed to work in the visible light regime (400–700 nm), the smallest features that are resolvable are about the size of half the wavelength of light used (approx. 200 nm). In order words, any two features that are closer than half the wavelength of the light would be regarded as a single feature. Thus optical microscopes have limited resolution in the nanoscale regime.

How small can an optical microscope resolve? The *Rayleigh criterion* is the commonly accepted criterion for the minimum resolution achievable. The imaging process is said to be diffraction-limited when the first diffraction minimum of the image of one point source coincides with the maximum of another (Fig. 8.2).

The mathematical expression of the empirical diffraction limit given by the Rayleigh criterion is:

$$\sin \theta = 1.22 \frac{\lambda}{D} \tag{8.1}$$

where θ is the angular resolution, λ is the wavelength of light, and D is the diameter of the lens. The factor 1.22 is derived from a calculation of the position of the first dark ring (Bessel function) surrounding the central Airy disc of the diffraction pattern.

The *resolution R* is defined as the minimum distance between distinguishable objects in an image. The resolution R depends on the angular aperture α:

$$R = \frac{1.22\lambda}{2n \sin \alpha} \tag{8.2}$$

where α is the collecting angle of the lens, which depends on the width of objective lens and its focal distance from the specimen.

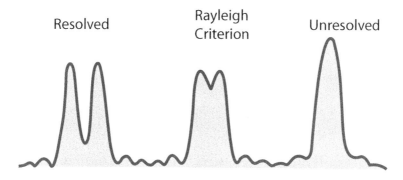

Figure 8.2. The Rayleigh criterion.

n is the refractive index of the medium in which the lens operates. λ is the wavelength of light illuminating or emanating from (in the case of fluorescence microscopy) the sample. The quantity $n \sin \alpha$ is also known as the numerical aperture. For the best lenses, α is about 70° ($\sin \alpha = 0.94$), the shortest wavelength of visible light is blue ($\lambda = 400$ nm), and the typical high resolution lenses are oil immersion lenses ($n = 1.5$). Substituting these values in Eq. (8.2), we determine R to be approximately 170 nm, i.e., the resolution limit of a light microscope using visible light is about 200 nm.

There is another consideration in developing optical microscopes with high magnification, namely, the light ray has to be focused very tightly onto the samples. This means such an optical microscope would have limited *depth of focus*. For example, a typical optical lens with a magnification of 100× would have a depth of focus of about 1-2 microns. As a result, if the object we are interested in seeing is a sizable 3D object (> a few microns thick), then we would only be able to obtain a sharp image of part of the object.

8.2 ELECTRON MICROSCOPY

In the early 1930's, many scientists and engineers realised that they have reached the theoretical limit of the resolving power of an optical microscope. In order to be able to "see" the finer details of objects such as biological cells, scientists started to develop new type of microscopes that make use of fast-moving electrons instead of light. Microscopes that make use of fast-moving electrons are known as electron microscopes and they are generally classified into two types: the Scanning Electron Microscope (SEM) and the Transmission Electron Microscope (TEM). The idea is to direct a focused beam of electrons towards a small part of an object in vacuum and detect various signals generated due to the interaction of the electrons with the object. Images can be generated depending on the contrast in the magnitude of the signals obtained when the beam of focused electrons is scanned across the object. The first electron microscope was invented by Ernst Ruska and Max Knoll from Germany. In 1986, the Physics Nobel Prize was co-awarded to Ernst Ruska for the development of electron microscopy.

8.2.1 Scanning Electron Microscopy

Efforts put into the development of electron microscopy have paid off as the electron microscopes can achieve better resolution and greater depth of focus. Why does an electron microscope achieve better resolution? Scanning electron microscope uses electrons with energies of a few thousand electron volts (eV). The de Broglie wavelength of an electron is given by $\lambda = h/p$, where h is Planck's constant and p is the momentum of the electron. For electrons with energy of 3600 eV, the wavelength is 0.02 nm. Hence the electron microscope would be able to achieve much better resolution than ordinary optical microscopes. Figure 8.3 shows some examples of SEM images.

In practice, the resolving power of ordinary electron microscopes is about one nanometre. This number is larger than the de Broglie wavelength because instrument geometry and electron scattering in the specimen are factors that influence the resolving power. Furthermore, during the operation of a scanning electron microscope, the profile of the focused beam of electrons

Figure 8.3. SEM images of (a) compound eyes of an ant, (b) compound eyes of a mosquito, (c) a strand of human hair and (d) the surface of the wings of a housefly (from author's lab).

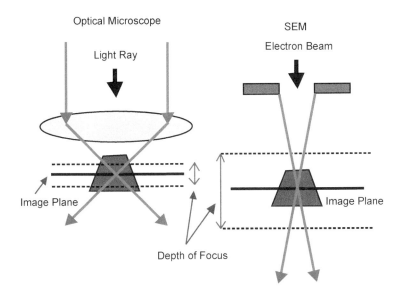

Figure 8.4. A comparison between the light beam profile in an optical microscope and electron beam profile in a SEM. The depth of focus of the optical microscope is shorter than that of the electron microscope.

is less converging and this give rise to a deeper depth of focus. As a result, the SEM is able to provide a sharper image of 3D objects as compared to an optical microscope. Figure 8.4 shows a schematic comparison that illustrates the difference in the depth of focus between an optical microscope and a SEM. The depth of focus of the optical microscope is shorter than that of the electron microscope. Having a larger depth of focus means that the electron microscope would be able to generate sharp images of the 3D object shown since the entire object is within the depth of focus.

How does a SEM work?

A SEM consists of an electron gun which produces the electrons, an applied high electric potential that accelerates the electrons, a system of electromagnetic lenses that focus the beam of electrons onto the sample, scanning coils that facilitate the scanning of the electron beam over the sample surface, the sample chamber where the sample is located, and detectors that measure the signals generated due to the interaction of the electrons with the sample. All these components are housed within a vacuum

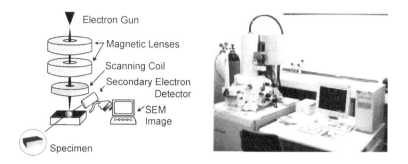

Figure 8.5. (a) Schematic drawing of the main components of a SEM. (b) Photo of a SEM unit.

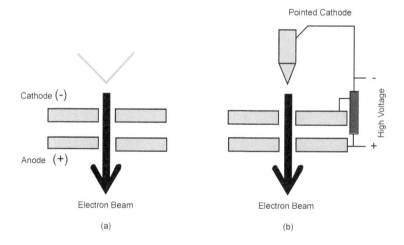

Figure 8.6. Schematic of operation of (a) a thermionic electron gun and (b) a field emission electron gun.

chamber. Figure 8.5 shows a schematic of a typical SEM and a photograph of a commercial SEM.

There are two types of electron source (or electron gun): thermionic or field emission guns. Figure 8.6 illustrates the operations of these two types of electron gun. A thermionic electron gun relies of electrons emitted from a heated wire or a filament. This filament is usually a bent tungsten wire that functions as the cathode. The bent portion becomes heated up when an electric current passes through the filament. The outer orbital electrons of the tungsten atoms are emitted when they gain sufficient thermal energy to overcome the energy barrier that prohibits the electrons from escaping. The higher the temperature, the higher the number of electrons emitted. Tungsten is commonly chosen as the filament because it can withstand high temperature without melting. However, thermionic electron guns have relatively low brightness.

On the other hand, the field emission electron gun relies on electrons emitted from a sharp tip upon the application of a high electric field. It does not involve heating of a filament. Instead, when a high electric field is applied to the tip, electrons from the tip quantum mechanically tunnel through the energy barrier into the vacuum. Typically the field emission gun has two anodes. The first anode (at $\sim$0–5 kV) serves to extract the electrons from the tip, while the second anodes (at $\sim$1–50 kV) serves to accelerate the electrons and this determines the energy of the electrons traveling down the column of the SEM. The field emission electron gun has a higher brightness.

As the electrons are streaming out from the electron guns, they form a spray pattern. In order to control the profile of this electron beam into a finely adjusted focused beam, electromagnetic lenses are used. When an electron with charge q and velocity $\vec{v}$ travels in a region with a magnetic field $\vec{B}$, it will experience a force $\vec{F}$ given by:

$$\vec{F} = q\vec{v} \times \vec{B} \tag{8.3}$$

One thing to note is that since the direction of the force is acting perpendicular to the direction of the velocity, the Lorentz force acting on the electrons has no effect on the speed of the electron. The only effect the magnetic field has on the electron is to change the direction of motion of the electrons.

As shown in Fig. 8.7, the magnetic field profile generated by a typical electromagnet used in a SEM is highly non-uniform. The magnetic field of the electromagnetic lens can be considered to be made up of two independent components, the vertical axial component (Hz) and the horizontal radial component (Hr). The radial component causes the electron traveling in the $-z$ direction to move in a helical manner with respect to the central axis. The axial component causes the electron to move closer to the central axis, i.e., the effect of the axial component is to reduce the diameter of the helical path of the electrons. As a result, the electron beam

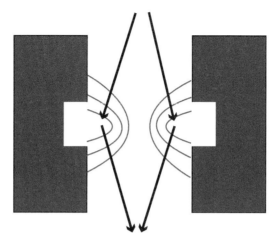

Figure 8.7. Schematic of the magnetic field profile generated by a typical electromagnet used in a SEM and the focusing effect of the magnetic field on the electron beam.

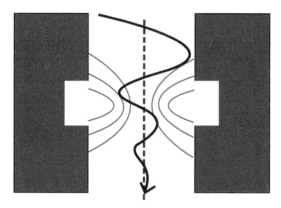

Figure 8.8. Spiral trajectory of an electron passing through the electromagnetic lens in a SEM.

spirals down the column as it passes through the electromagnetic lens as shown in Fig. 8.8. The resultant effect is that the electron beam becomes finely focused and can be scanned over the sample for imaging purposes.

There are usually two such electromagnetic lenses in the SEM system, the condenser lens and the objective lens. The condenser lens provides the first focusing effect and guides the electrons traveling through the SEM column. The objective lens focuses the electron beam onto the sample surface. The focal length of the objective lens is denoted as the working distance of the microscope.

The scanning of the electron beam over the surface of the sample is achieved by deflecting the beam using an applied electric field or magnetic field. Typically a deflection coil consists of four radially oriented coils arranged so that the magnetic field is perpendicular to the axis of the system. The magnetic field generated by these coils can be controlled by the amount of electric current passing through these coils. By programming these scanning coils, one can readily raster the electron beam over the sample surface.

The typical accelerating voltage used in a SEM is of the order of a few thousand volts. With an energetic beam of electrons scanning over the sample surface, a number of phenomena occur due to the interaction between the electrons and sample atoms. The incident electrons can collide with the electrons of the atoms, or they can collide with the atomic nuclei. Figure 8.9 illustrates the wealth of phenomena that are observed when an energetic electron beam is incident on a typical sample.

We next briefly describe the nature of the detectable signals and their applications in a SEM. Energetic incident electrons can

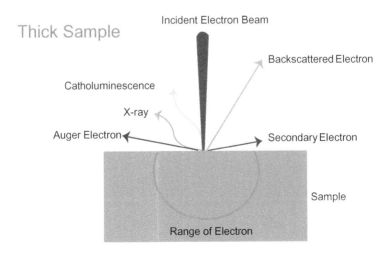

Figure 8.9. Detectable signals generated when an energetic electron beam is incident on a typical sample.

collide with the electrons in the sample and knock them out of their usual orbits. These electrons are known as Secondary Electrons (SE). During the process, the incident electron loses little energy and continues to generate more SE as it travels further into the sample. A single incident electron will typically generate a shower of thousands of SE until the incident electron loses its energy. Since a large number of SE are generated, the detection of SE is the most common mode of operation for SEM sample imaging. Note that the SE have low energies so that SE generated deep in the sample are unable to travel to the surface and leave the sample. As a result, the SE detected are primarily from region close to the sample surface (<10 nm). Hence SE imaging would produces good topographical information of the sample.

Sometimes an incident electron collides with the nucleus of a sample atom, causing the electron to bounce back. Such electrons are referred to as Backscattered Electrons (BE). Since the atomic nucleus is more massive than the electron, the BE has high velocity and is characterized by its high energy (a few keV). High density samples will generally create more BE, and hence BE imaging can be utilized to identify differences in the densities of a sample. The production of BE varies directly with the atomic number of the atoms in the sample. Therefore regions with atoms of higher atomic number would appear brighter than regions with atoms of lower atomic number (see Fig. 8.10). In this way, besides providing information on the topography of the sample, detection

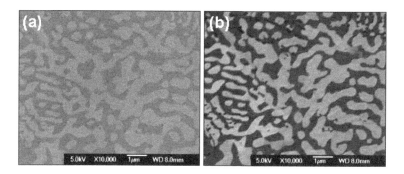

Figure 8.10. (a) SEM image of solder using secondary electron imaging mode. (b) SEM image of the same sample obtained using the backscattered electron (BE) imaging mode. Brighter areas in the BE image correspond to lead phase of solder (from author's lab).

of BE provides valuable information on the density and elements distribution in the sample.

Sometimes the incident electron collides and removes an electron from the inner shell the sample atom, leaving a hole in the orbital. An electron from a higher orbital will make a transition to this vacant lower energy level, filling the vacancy. During this transition, the difference in the energy is emitted in the form of electromagnetic radiation. Typically this radiation falls in the X-ray regime. The X-ray photons emitted in these processes are unique to each element and can be used to identify the elements in the sample. This technique requires an X-ray detector, and is known as energy dispersive X-ray spectroscopy (EDX), typical spectra of which are shown in Fig. 8.11. There is another competing process to X-ray emission — the emission of an Auger electron (AE). Instead of releasing the difference in energy in the form of a X-ray photon, this energy can be transferred to an electron occupying another outer shell, which leaves the sample. Such an electron is known as the Auger Electron

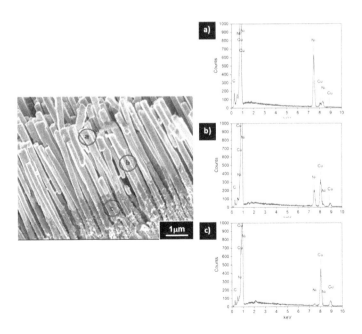

Figure 8.11. Example of SEM image of hybrid nanorods and EDX spectra obtained fron different parts of the nanorods (from author's lab).

(AE). The AE has a characteristic energy, unique to each element from which it was emitted. AE has relatively low energy and are only emitted from the surface of the specimen, typically from a depth of <3 nm, thereby yielding surface sensitive compositional information. Hence, both EDX and Auger electron spectroscopy are used for elemental analysis or chemical characterisation of a sample.

Light is also emitted when a sample is being bombarded with electrons. Many substances give out light when bombarded with electrons, just like a TV monitor. This effect can be exploited for imaging. The light emitted can be in the ultra-violet, visible or infrared range and this phenomenon is referred to as Catholuminescence (CL).

The SEM generates an image of the sample by scanning the electrons over the sample surface while a SE detector placed near the sample collects the signal generated. Modern SEMs have incorporated many attractive technical features so that imaging with a SEM has become very user friendly. The images are generated almost real time and high quality images can be stored directly in digital format. A simple turn of a knob allows us to

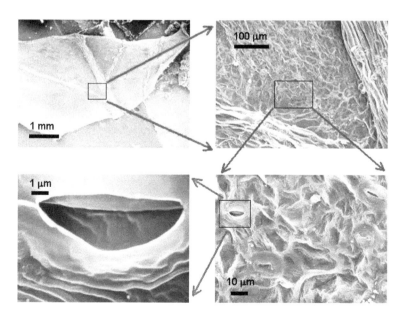

Figure 8.12. SEM images of a leaf sample at various stages of magnification (from author's lab).

change the magnification of the system and zoom into the sample for a close-up view. Figure 8.12 shows sequential images of the same sample at different magnifications. In an optical microscope, the notion of higher magnification corresponds to a more tightly focused light cone. In the SEM, operating at a higher magnification means scanning the focused electron beam over a smaller area of the sample. It is remarkable that a typical SEM can achieve a wide range of magnification from $25\times$ to $600,000\times$. Naturally when the magnification reaches a high value, more effort is required to obtain a high quality image. The quality of the image also depends on the type of sample being imaged too.

The SEM requires a vacuum environment in order to function properly. If a poor vacuum is maintained in the system, air molecules can cause the electron source to burn out. The electron beam would also be scattered by the air molecules in the chamber. The collision between the electrons and the air molecules could give rise to ionization and discharge. The stability of the beam and the quality of the images would be affected. The presence of air molecules in the SEM system can result in chemical reactions between the sample and the molecules. The result is the formation of some compound on the sample. This will affect the quality of the image too.

8.2.2 *Transmission Electron Microscopy*

Historically, the development of the electron microscope actually started with the development of the Transmission Electron Microscope (TEM). As the name suggests, during the operation of the TEM, the electrons pass through a sample. Naturally the sample is required to be very thin, and there are specialised methods for sample preparation. Bulk materials are thinned to make them electron transparent by simply crushing them and depositing some fragments on a carbon foil, or by mechanical grinding and ion milling. Nanoparticles are thin enough for direct observation by typically depositing them on a conducting sample grid.

In TEM, the voltage used to accelerate the electron is much higher than that used in the SEM, typically in the range of 200 to 300 kV. At such high energies, the electrons are able to pass through a thin sample. The de Broglie wavelength of such energetic energy electrons would be very short, and this means the TEM is able to image even smaller features than the SEM.

High Resolution TEM (HRTEM) is now routinely used to achieve atomic resolution of a sample. However, TEM has its limitations. Lengthy sample preparation is usually required to make the sample thin enough. Since the beam is traveling through the sample, the sample bulk and not the surface is being imaged.

How does a TEM work?

The working principle of a TEM is very similar to that of a SEM. Figure 8.13 shows a schematic and photograph of a typical TEM. Similar to SEM, an electron gun that produces a stream of monochromatic electrons is typically located at the top of the instrument. This stream of electrons is focused to a coherent beam by condenser lenses 1 and 2. Condenser aperture is employed to restrict the beam and remove high angle electrons that deviate from the main optic axis of the system (indicated by the dotted line). The beam of electrons strikes the specimen and parts of it are transmitted. The transmitted electrons are then focused by the objective lens into an image. The objective and selected area metal apertures are utilised depending on the mode of imaging. The two basic operations of the TEM imaging system are the image projection and diffraction pattern projection. During the image

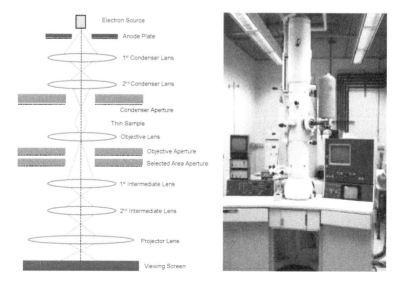

Figure 8.13. Schematic of a typical TEM setup and photograph of a TEM unit.

projection mode, the objective aperture is engaged and it enhances contrast by blocking high-angle diffracted electrons. During the diffraction pattern projection mode, the selected area aperture is engaged and it facilitates the examination of the periodic diffraction of electrons caused by the interaction of the electrons with the ordered arrangements of atoms in the sample. The formed image then passes along the TEM column through the intermediate and projector lenses before it strikes the fluorescent screen. In the image projection mode, the darker areas of the image represent those areas of the sample that fewer electrons are transmitted through (thicker or denser); the lighter areas represent those areas that more electrons are transmitted through (thinner or less dense).

When the incident electrons strike the sample, the usual phenomena found in a SEM such as SE, BE, AE, X-ray, CL are generated as previously discussed. However, in the case of a TEM, since the sample is thin, electrons pass through the sample. We classify the electrons that pass through the sample into three main categories, namely, the unscattered electrons, the elastically scattered electrons and the inelastically scattered electrons (see Fig. 8.14).

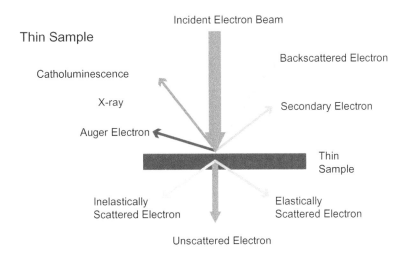

Figure 8.14. Schematic of the signals generated when an incident electron beam strikes a thin sample in a TEM.

The unscattered electrons are incident electrons that pass right through the thin sample without any interaction with the sample atoms. The transmission of such unscattered electrons is inversely proportional to the thickness of the sample: The thicker the sample, the fewer the transmitted electrons and vice versa. Therefore the thicker parts of the sample appear darker while the thinner parts of the sample appear lighter. Figure 8.15 shows TEM images of various nanostructured materials, and Figure 8.16 shows atomically resolved lattice fringes of selected nanomaterials.

When the incident electrons are deflected due to their interaction with the sample but with no loss of energy, this gives rise to elastically scattered electrons. During the elastic interaction, incident electrons are scattered according to Bragg's law of diffraction by atoms with regular atomic spacing. The scattered electrons are collated using magnetic lenses to form a pattern of diffraction spots. These diffraction patterns can be used to identify the crystalline order of the sample, giving information

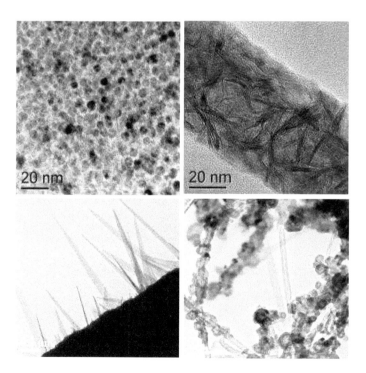

Figure 8.15.   Examples of TEM images (from author's lab).

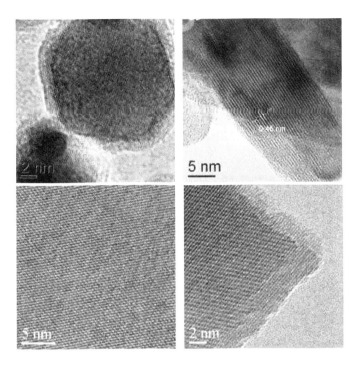

Figure 8.16. Examples of high resolution TEM images (from author's lab).

about the orientation, and atomic arrangement in the area probed. This mode of operation is known as Selected Area Electron Diffraction (SAED). Figure 8.17 shows SAED patterns of various nanomaterials.

Inelastically scattered electrons are generated when the incident electrons lose energy due to their interactions with the sample atoms. The inelastic loss of energy by the incident electrons is characteristic of the elements that were interacted with. These energies are unique to the bonding state of each element and thus can be used to extract both compositional and bonding information of the specimen region being examined.

8.3 SCANNING PROBE MICROSCOPY

Optical microscopes make use of light to achieve high magnification of the sample, while electron microscopes make use of

Figure 8.17. Examples of SAED images (from author's lab).

electrons to allow us to probe even further into the atomic structure of materials. Scanning Probe Microscopes (SPM) represent yet another class of microscopes that allows us to acquire very high magnification images of samples. This class of microscopes works using a totally different principle. Imagine being in a room that is completely pitch dark so nothing is visible. Our instinct in navigating in such a room would be to reach out our hands and feel our way around the room. SPMs work in a very similar manner whereby the world of small tiny objects is revealed by "feeling" around the surface of a sample by a sharp sensor. Depending on the type of sensors employed, we have different types of SPM. In the following two sections, we shall describe the two most commonly used SPMs: the Scanning Tunneling Microscope (STM) and the Atomic Force Microscope (AFM).

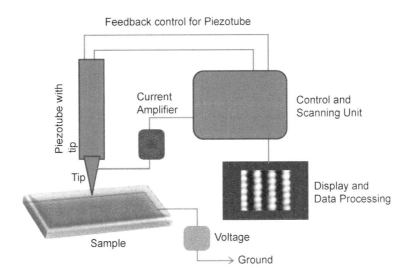

Figure 8.18. Schematic diagram of a Scanning Tunneling Microscope.

8.3.1 *Scanning Tunneling Microscopy*

The STM was invented by Gerd Binnig and Heinrich Rohrer from IBM Zurich Lab in Switzerland in 1981. In their seminal paper,[1] they described the construction of the first STM where a sharp needle was used to probe the surface of a piece of silicon. The instrument was able to achieve atomic resolution and revealed the ordered arrangement of Si atoms on the surface of the sample. This outstanding work won them the 1986 Physics Nobel Prize. The invention of STM opened a new window to the world of nanoscience and nanotechnology, whereby surfaces can be imaged atom-by-atom in real space. STM has since developed into an important tool in physics, chemistry, biology, engineering and materials science.

The working principle behind the operation of the STM is quantum tunneling (cf. Chapter 3). Figure 8.18 shows the schematic of a typical setup for a STM. The most critical component of a STM is the sharp tip used in imaging the sample surface. Usually the sharp tip is prepared so that there is a single atom that protrudes at the tip. When the tip is brought very close to surface

---

[1] G. Binnig, H. Rohrer, Ch. Gerber and E. Weibel, *Appl. Phys. Lett.* **40**, 178–180 (1982).

of the sample with proper biasing, electrons will tunnel across the gap between the atom at the tip and the atom of sample located directly underneath the tip. The magnitude of the tunneling current (I) depends exponentially on the distance (d) between the probing atom and the sample.

$$I \propto \exp\left(-Kd\right) \tag{8.4}$$

where K is the characteristic exponential inverse decay length.

Since only the atom closest to the probing atom will contribute significantly to the detected current, STM is able to achieve atomic resolution. As the magnitude of the tunneling current is small, the tunneling current is amplified by an amplifier. To scan the tip across the surface of the sample, the tip is attached to a piezo-electric tube with controlling electrodes. The piezoelectric tube is made of piezoelectric crystal. Application of an electric field causes a strain resulting in the deformation of the crystal. As a result, the tip that is attached to the tube will be displaced by a small amount. The magnitude of the deformation, i.e. the movement of the scanning tip, depends on the electric field applied. Hence the scanning of the tip can be precisely controlled. Since we are making use of the tunneling current between the sample and the tip, the sample has to be conducting. A control unit controls the feedback to the piezoelectric tube and also captures the tunneling current detected. The data is processed and displayed on a computer monitor.

The sensor tip of the STM is an important component of the microscope. Such sharp metallic tips are typically prepared by electrochemical etching. Movement of the tip in XYZ directions with sub-angstrom accuracy is controlled via the piezoelectric rods. For coarse movement, the entire scanning assembly is positioned using micro-motorised platform. Since the magnitude of the tunneling current is small, an amplifier is required in the feedback loop and computer-based data collection system. To reduce the noise from mechanical vibrations, the STM is housed in a platform with vibration isolation.

The STM can operate both in ambient or in a controlled environment such as in a vacuum system. More specialised STMs can operate in liquid media. When the best atomic resolution is required, the STM operates in an ultrahigh vacuum (UHV) environment, and the system usually incorporates a deposition

Figure 8.19. Left: Photograph of an ultrahigh vacuum (UHV) STM system; Right: Close-up of STM sample stage and tip (from author's lab).

chamber and other characterisation capabilities as well as the STM. A typical UHV STM system is shown in Fig. 8.19, whereby controlled atomic layer deposition can be performed and sample contamination prevented since the sample needs not be exposed to air.

Example 1: Si(111)-(7 × 7) reconstruction

The very first surface structure imaged by Binnig and Rohrer was the Si(111)-(7 × 7) reconstruction. The Si(111)-(7 × 7) is a complex but intriguing surface that has been extensively studied by surface scientists. Figure 8.20(a) shows the details of the unit cell of the structure and Fig. 8.20(b) shows an STM image obtained for such a surface. Each bright spot in the image represents regions on the sample surface with high densities of tunneling electrons. At this particular bias, the bright spots correspond to the location of individual Si atoms on the top layer of silicon. This image demonstrates the superior atomic resolving power of the STM which can clearly resolve individual atoms as shown.

Example 2: Silicon carbide 6H-SiC(0001) reconstructions

The discovery of graphene (single 2D layer of graphite) has opened up a new paradigm in nanoelectronics that could offer

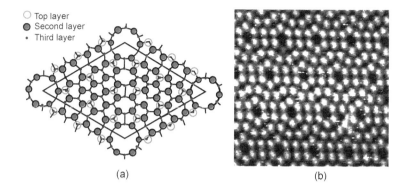

Figure 8.20. (a) Structure of the Si(111)-(7 × 7) reconstruction. (b) Typical STM images of Si(111)-(7 × 7) (image (b) is from author's lab).

better performance than conventional semiconductor devices owing to its unusual Dirac fermion behaviour of its electrons that gives rise to superior mobility and unique anomalous quantum Hall effect.[2] The thermal decomposition of silicon carbide (SiC) is an attractive route to grow epitaxial graphene as it provides the most direct route for integration and tailoring its properties with minimum modification to existing SiC technology. Silicon-rich 6H-SiC(0001) undergoes a series of surface reconstructions with increasing annealing temperature, from the Si-rich 3×3 reconstruction to the C-rich $6\sqrt{3} \times 6\sqrt{3}$-R30° "nanomesh", before the 1×1 graphene phase is finally formed.[3,4] Figures 8.21(a)–(c) show the low energy electron diffraction (LEED) patterns and corresponding STM images obtained for SiC surface as it is annealed to increasing temperatures. Understanding the graphene growth process is fundamentally important as it allows us to manipulate and control the transformation process and possibly tune the band gap of epitaxial graphene on SiC.

[2] K. S. Novoselov, A. K. Geim, S. V. Morozov, D. Jiang, M. I. Katsnelson, I. V. Grigorieva, S. V. Dubonos and A. A. Firsov, *Nature* **438**, 197 (2005).
[3] W. Chen, H. Xu, L. Liu, X. Y. Gao, D. C. Qi, G. W. Peng, S. C. Tan, Y. P. Feng, K. P. Loh and A. T. S. Wee, *Surface Science* **596**, 176 (2005).
[4] S. W. Poon, W. Chen, E. S. Tok and A. T. S. Wee, *Appl. Phys. Lett.* **92**, 104102 (2008).

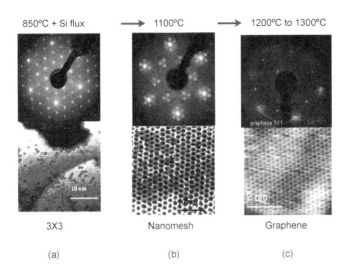

Figure 8.21. Low energy electron diffraction (LEED) patterns (on top) and the corresponding STM images of (a) 3×3, (b) $6\sqrt{3} \times 6\sqrt{3}$-R30° "nanomesh", and (c) 1×1 graphene reconstructions of 6H-SiC(0001) after different annealing conditions (from author's lab).

Example 3: Molecular self-assembly on surfaces

STM can also be used to image molecules on surfaces, and this is typically carried out in UHV and at low temperatures (77 K or 4 K) to minimize molecular diffusion and thermal effects. C_{60} fullerene molecular chains have been shown to form neat 1D chains on self-assembled α-sexithiophene (6T) molecules on Ag(111) substrates (Fig. 8.22).[5]

Modes of Operation for STM

The STM can operate in two modes: constant-height mode and constant-current mode. Let us denote the sample surface as the XY plane and the vertical motion of the tip as the Z direction. In the constant height mode, the STM tip is scanned back and forth across the surface with the Z-position constant throughout (Fig. 8.23(a)). As the tip traverses across the surface, the tunneling current is recorded. The computer converts the tunneling current

---

[5] H. L. Zhang, W. Chen, L. Chen, H. Huang, X. S. Wang, J. Yuhara and A. T. S. Wee, *Small* **3**, 2015 (2007).

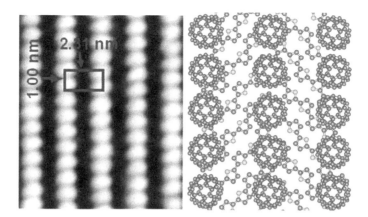

Figure 8.22. STM image (left) of C_{60} molecular chains on self-assembled α-sexithiophene (6T) molecules on Ag(111). Right: Corresponding molecular model (from author's lab).

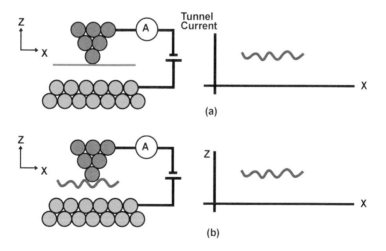

Figure 8.23. Illustration of the (a) constant-height and (b) constant-current modes of operation.

into a line plot or image reflecting the magnitude of the tunneling current. In constant-current mode, the STM tip is rastered across the XY plane with continuous adjustment of its Z-position such that the tunneling current remains the same via a current feedback loop (Fig. 8.23(b)). In this case, the topographical contour of the sample surface is generated by plotting the variation of the Z position of the tip as it is scanned across the surface. This mode of operation is usually preferred as it prevents the tip from crashing into the surface.

During the STM experiment, when the tip is brought near the sample surface and the tunneling current is recorded, does it matter which way the tunneling electrons flow? How does the tunneling direction depend on the voltage bias applied to the system? To answer these questions, it is helpful to look at the energy band diagram for the tip-sample system. Figure 8.24(a) shows the energy band diagram for the STM tip and the sample separated by a small gap (vacuum or air barrier) without any voltage bias applied to the sample or the tip. Both the energy bands are typical of a conductor where the electrons fill the energy levels up to the Fermi level, according to Pauli's exclusion principle. Without any voltage bias, the Fermi levels in the tip and sample are aligned, and there is no net electron-tunneling across the vacuum gap. Practical operation of the STM requires the application of a voltage bias across the tip and sample. When the STM tip is negatively biased (magnitude of the voltage bias is V) relative to the sample as depicted in Fig. 8.24(b), the energy levels of the tip is raised by an amount eV with respect to the energy levels of the sample. Hence, electrons from the tip within the band of eV from the Fermi level readily tunnel across the gap into the sample. On the other hand, when the STM tip is positively biased relative to the sample, the reverse situation occurs as illustrated in Fig. 8.24(c), and electrons tunnel from the filled states in the sample to empty states in the tip. Thus the direction of flow of the tunneling electrons depends on the voltage bias adopted during the experiment. In addition, the magnitude of the measured current depends on the magnitude of the applied voltage bias.

After the STM tip has completed imaging the sample surface under bias conditions, the computer program generates a false color image with little dots. Does this STM image represent the real positions of the individual atoms on the surface? It turns out that the answer to this question depends on the sample under investigation. Strictly speaking, the STM image represents the spatial variation of the electronic density at the surface. We may be "seeing" the atoms in some images, but not in others. We shall discuss a typical case of a semiconductor where care has to be taken in interpreting the image.

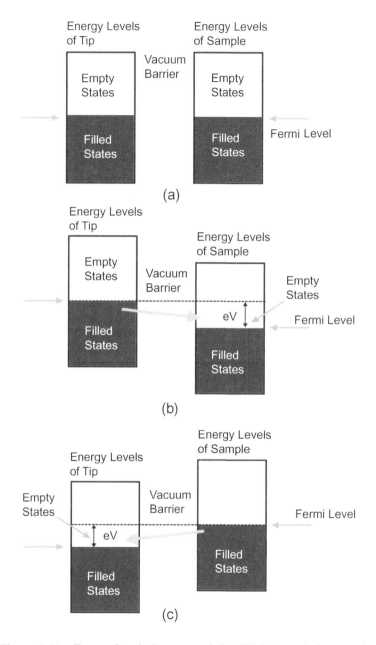

Figure 8.24. Energy band diagrams of the STM tip and the sample (a) without any voltage bias, (b) when the tip is negatively biased with respect to the sample, (c) when the tip is positively biased with respect to the sample.

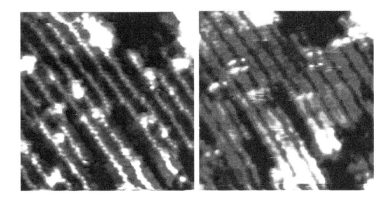

Figure 8.25. 28 nm × 28 nm STM scans of filled (left) and empty (right) state images of GaAs(001) surface in the same scanning region. The left image shows single As dimer chains, and the right image shows double Ga dimer chains in the corresponding trough between the As chains (from author's lab).

Case study: GaAs(001)

Semiconducting surfaces with localised bonds typically show strong bias-dependent images. Figure 8.25 shows the STM images of Ga-dimers on a GaAs(001)-c(8 × 2)-Ga surface, where it can be clearly seen that different bias conditions reveal different images.[6]

Besides being a powerful tool that provides high-resolution images of the arrangement of atoms of the surface of a sample, STM can be utilised to provide additional valuable information about the electronic states of the atoms. The operation of the STM can be readily adjusted to perform the Scanning Tunneling Spectroscopy (STS) studies of the sample atoms. In the simplest STS mode, the tip is placed above a specific site on the sample surface. Without changing the distance between the tip and the sample, the bias voltage is continually varied while the tunneling current is continually measured. From the resulting I-V measurements, details on the energy levels of the sample and its density of state can be determined.

---

[6] H. Xu, Y. Y. Sun, Y. G. Li, Y. P. Feng and A. T. S. Wee and A. C. H. Huan, *Physical Review B* **70**, 081313 (2004).

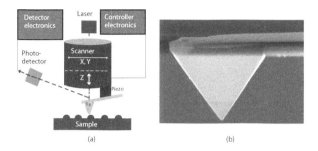

Figure 8.26. (a) Schematic of a typical AFM. (b) Image of the cantilever and probe tip (from author's lab).

8.3.2 *Atomic Force Microscopy*

The Atomic Force Microscope (AFM), or Scanning Force Microscope (SFM) was invented in 1986 by Binnig, Quate and Gerber.[7] The AFM comprises a sharp tip at the end of a cantilever which bends in response to the force between the tip and the sample. As the cantilever with the sharp tip is scanned across a sample, the deflection of the cantilever is detected. Hence, as the cantilever is scanned across the sample, the detected deflections would be utilised to generate a map of surface topography of the sample. Figure 8.26(a) shows a schematic of a typical setup of the AFM.

As shown in Fig. 8.26(b), the AFM cantilever is typically about 100 to 200 microns in length and the probe tip is located at the free end of the cantilever. The interaction force between the tip and the sample causes the deflection of the cantilever. Once the deflection is measured, the magnitude of the force experienced can be determined by the product of the spring constant of the cantilever and the deflection. As the AFM relies on inter-atomic forces for its operation, it can be used to study insulators and semiconductors as well as electrical conductors.

It should be noted that in a typical AFM scan, the cantilever exhibits a tiny deflection. For example, with a force of 1 nN and a spring constant of 0.5 N/m, the deflection of the cantilever would be only 2 nm. Hence sensitive schemes need to be developed to detect the deflection of the cantilever. Figure 8.27 illustrates various modes of detection for the deflection of the cantilever.

---

[7] G. Binnig, C. F. Quate and Ch. Gerber, *Phys. Rev. Lett.* **56**, 930 (1986).

Optical detection as illustrated in Fig. 8.27(a) is the most commonly adopted method to measure the deflection of the cantilever. In this method, a fine laser spot is focused onto the back of a cantilever. The back of the cantilever is typically coated with a thin layer of gold to improve its reflectivity. A position sensitive photon-detector is used to detect the position of the reflected spot. As the cantilever bends, it causes the position of the reflected spot to change and this gives rise to a voltage change in the photon-detector. Figure 8.27(b) illustrates another detection mode whereby a STM tip is positioned very close to the back of the cantilever to measure the tunneling current between the tip and the cantilever. If the cantilever bends, the tunneling current changes with tip-cantilever separation. Since the tunneling current is a sensitive function of the distance between the tip and cantilever, the deflection of the cantilever can be accurately determined.

Specially made cantilevers where the cantilever material is piezoresistive (Fig. 8.27(c)) have been utilised. Bending of the cantilever will cause strain in the material and its resistance will change. Hence by measuring the resistance, one can tell by how much the cantilever has bent and thus the deflection is measured. In another mode of detection, changes in the capacitance can be employed to detect the deflection of the

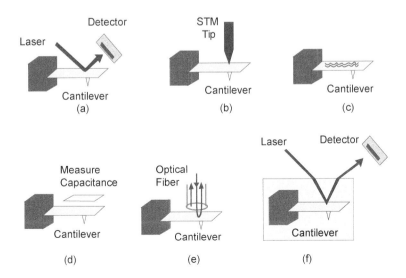

Figure 8.27. Illustrations of the various detection modes to measure the deflection of the AFM cantilever.

cantilever (cf. Fig. 8.27(d)). The back of the cantilever and a parallel plate forms a parallel-plate capacitor, which is sensitive to the separation between the two plates. Figure 8.27(e) shows a detection mode that makes use of optical interferometry. Here, an optical fibre carries laser light that shines on the back of the cantilever. The laser light is reflected at two locations, on the back of the cantilever and the end wall of the optical fibre. These two reflected beams give rise to an interference pattern that depends on the distance between the optical fibre and cantilever, hence providing a measurement of its deflection.

The AFM has been widely used in many disciplines since the technique is applicable to different types of samples, conducting or not, in liquid or in air. One of the main advantages of the AFM is that it can be used in an aqueous environment, making it particularly useful in biology. Figure 8.27(f) shows the schematic of AFM operation in an aqueous medium. In this case, the optical detection mode is preferred since the laser beam can readily pass through the transparent medium. Using such a setup, one can obtain images of biological cells in aqueous medium. Examples of some AFM images obtained are shown in Fig. 8.28.

Since the operation of the AFM relies on the interatomic interactions between the probing tip and the sample surface, there are different force regimes for the operation of the AFM. This is typically classified into the contact and non-contact modes of operation. In the contact mode, the cantilever is positioned very close ($<$ a few angstroms) from the sample surface. At this range, the interatomic force between the cantilever and sample is repulsive, and the magnitude of the force varies from 10^{-8} N to 10^{-6} N. In the non-contact mode, the cantilever is positioned at about 10 to 100 angstroms from the sample surface and measures sample topography with little or no contact between tip and sample. At this range, the interatomic force between the cantilever and the sample is attractive with a typical force magnitude of 10^{-12} N. In the non-contact mode, the cantilever is oscillated near its resonant frequency (typically 100 to 400 kHz) with an amplitude of a few tens of angstroms. Changes to the resonant frequency or vibration amplitude are detected as the tip approaches the sample surface. Since the force required are small in the non-contact mode, it is well-suited for studies of soft or elastic samples such as biological cells or DNA molecules.

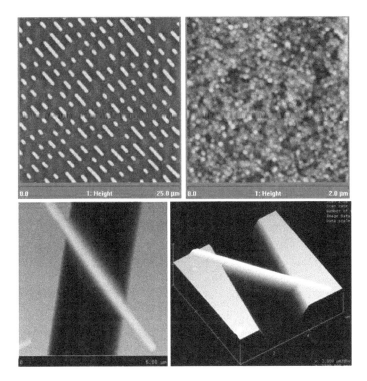

Figure 8.28. Examples of some images obtained via AFM (from author's lab).

8.4 OPTICAL TWEEZERS

First developed by Art Ashkin and co-workers in 1986,[8] optical tweezers refer to the technique where a tightly focused laser beam is used to trap tiny particles whose refractive index is greater than that of the surrounding medium. Such optical tweezers have been extensively employed in the manipulation of tiny objects such as microspheres made of polystyrene or silica, in studies of microscopic interactions, in biophysics, in the assembly of micro-beads, and in the creation of micron-sized structures.

The main feature of an optical tweezers setup is the creation of a tightly focused laser beam. This is achieved by introducing a parallel beam into an optical microscope. In this way the objective lens of the optical microscope can be utilized to create a tightly

---

[8] A. Ashkin, J. M. Dziedzic, J. E. Bjorkholm and S. Chu, *Opt. Lett.* **11**, 288 (1986).

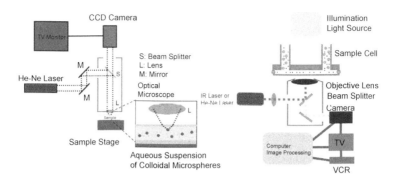

Using Upright Optical Microscope   **Using Inverted Optical Microscope**

Figure 8.29. Schematics of typical setups employed for optical tweezing.

focused laser beam. In addition, the same objective lens can be used to capture images of the microscopic particles trapped by the optical tweezers. Figure 8.29 shows schematics of two typical setups for optical tweezing. Both an upright or inverted optical microscope can be utilised. The housing of these optical microscopes consists of a side port that serves as the entrance point for the parallel laser beam. Inside the optical microscope, the laser beam is reflected by a beam splitter towards the objective lens to achieve focusing. The sample chamber is made of transparent housing that supports an aqueous colloidal suspension of micron-sized objects. The sample chamber is placed on the sample stage of the optical microscope. When the tightly focused laser beam is present in the colloidal suspension, the micro-particles are readily attracted towards the focused laser spot and become trapped.

When an intense focused laser beam is illuminated onto a micro-particle, the force exerted by the laser beam onto the particle can be divided into two main components: the scattering force and the gradient force. The scattering force arises from the change in momentum when a photon scatters off from the particle. This force tends to push the particle along the direction of beam propagation. On the other hand, the gradient force provides the attractive force component that draws the particles towards the focal point of the laser beam. At equilibrium the particle is held at a location slightly beyond the focal point of the focused laser beam.

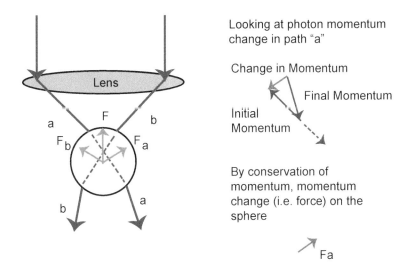

Figure 8.30. Laser beam profile passing through the microsphere.

The origin of the gradient force is explained as follows. In the Mie regime, the size of the particle is much larger than the wavelength of the laser beam. One can trace the laser beam path by applying the Snell's Law of refraction given the refractive indices of the medium and the particle. Figure 8.30 illustrates the case where a microsphere is positioned lower than the focal point of the laser beam. After passing through the microsphere, the profile of the laser beam gives additional momentum in the downward direction, and the recoil of the microsphere pushes it upward towards the focal point of the laser beam. Detailed analysis can be carried out and the general outcome is that the interaction between the laser beam and the microsphere always results in a force drawing the microsphere towards the focal point of the laser beam. In the other limit, the Rayleigh regime, the size of the particle is much smaller than the wavelength of light. Here the particle is regarded as a dielectric material in the electric field of the laser electromagnetic wave. The electric field induces a dipole moment in the particle. This interaction results in a strong trapping force for the particle located near the focal point of the beam profile.

There are a few experimental techniques employed to calibrated the trapping force exerted by the optical tweezers on a microsphere. One technique involves flowing fluid past a trapped

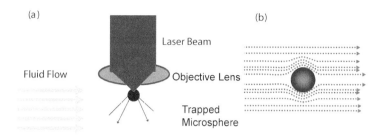

Figure 8.31. Force calibration of optical trapping force by fluid drag force.

microsphere as shown in Figure 8.31. The sphere in the fluid flow will experience a viscous drag force F, given by:

$$\text{Drag Force F} = 6\pi R\eta v \tag{8.5}$$

where R = radius of sphere, η = viscosity of water = 0.001002 Ns/m^2, v = velocity of the flowing fluid. Initially at low flow velocity, the sphere remains trapped because the optical trapping force is stronger than the drag force. As the flow velocity increases, the drag force increases and causes a slight deviation of the bead from its equilibrium position. At flow velocity greater than a critical velocity, known as the terminal velocity, the drag force becomes greater than the maximum optical trapping force and the sphere becomes detached from the optical tweezers. Thus the maximum optical trapping force achievable at a fixed laser power corresponds to the viscous drag force at terminal velocity. Typically, the force exerted by the focused laser beam falls in the range from a few pN up to a few hundred pN depending on the power of the laser beam employed.

Another force calibration technique makes use of a video-tracking method or a position sensitive detector to accurately determine the position of a trapped microsphere. The fluctuation in the position of the trapped microsphere due to thermal fluctuation can be captured, and the extent of fluctuation gives a measure of the stiffness of the optical trap. In order to measure the trap stiffness α, the position of the beads within the trap must be measured to nanometre or better resolution. Once the positions of the beads are accurately determined, one can make use of equipartition theory to determine the trapping strength of the optical tweezers. Accordingly,

$$\frac{1}{2}k_B T = \frac{1}{2}\alpha\langle x^2\rangle \tag{8.6}$$

where x is the displacement of the particle from its trapped equilibrium position, α is the stiffness, k_B is the Boltzmann constant and T is the temperature of the system.

Since the advent of the optical tweezers, different methods have been employed to shape the laser beam into different configurations for the purpose of optical trapping of assemblies of colloidal particles. These optical traps provide an important tool in the mesoscopic environment to trap and manipulate microscopic objects. Patterned laser beams have also been developed for many purposes such as the assembly of polymerised colloidal structure into micro-fluidic devices, driving forces for micro-devices and for cell-manipulation. One commonly adopted technique to create an array of laser spots is to use a diffractive optical element. Other techniques include the use of galvanometer scanning mirrors or a piezoelectric scanning mirror, computer-generated holograms and acoustically modulated light beams. Examples of optical trapping of an array of microspheres are shown in Fig. 8.32. Besides microspheres, other micron-sized objects such as biological cells and CuS micro-stars can be readily manipulated.

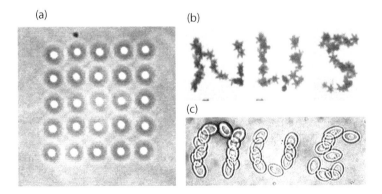

(a) (b) (c)

Figure 8.32. (a) Optical trapping of a 5×5 array of microspheres. Optical manipulation of the (b) CuS microstars and (c) fish blood cells into "NUS" letters formation (from author's lab).

Laser Nanofabrication

Carbon nanotubes (CNTs) have been of great interest to researchers due to their remarkable structural, electronic and mechanical properties. As they have tunable electronic properties, they are being investigated as possible new materials for the next generation of microelectronics and nanoelectronics devices. They have also been identified as potentially useful materials for a broad range of functions including actuators, fuel cells, flat panel displays, biosensors and improved X-ray sources. Arrays of CNTs have also been used as templates for a variety of materials to achieve functions such as creating super-hydrophobic surfaces. Many of these applications require a fabrication method capable of producing extended areas of patterned and aligned CNTs with uniform structures and periodic arrangements to meet device requirements. There are a few conventional methods to fabricate patterned aligned CNTs arrays. One such technique makes use of electron beam lithography to define the patterns of catalytic nanoparticles for the growth of CNTs.

In this section, we introduce a simple technique that makes use of a tightly focused laser beam from a moderate power laser as a precision cutting tool to create unique three dimensional 3D CNTs structures in ambient.[9] Figure 8.33(a) shows a schematic of the experimental set-up of an optical microscope-focused laser beam system in the author's lab. The system consists of an optical microscope and a medium power laser such as a He-Ne laser source ($\sim$20–40 mW). The role of microscope is to focus the laser beam onto the sample and at the same time capture the image of the sample. A parallel beam from the laser is directed into the microscope via two reflecting mirrors (M). Inside the microscope, the laser beam is directed towards an objective lens (L) via a beam splitter (S). The laser beam is then focused by the objective lens (L) onto the CNTs. Typically, an objective lens with high magnification and long working distance is preferred. When the focused laser beam is incident on the CNT sample, it readily trims away the top layers of the CNTs. SEM analysis reveals that part of the CNTs are removed cleanly without leaving behind any residue.

---

[9] K. Y. Lim, C. H. Sow, J. Y. Lim, F. C. Cheong, Z. Y. Shen, J. T. L. Chin, K. C. Chin and A. T. S. Wee, *Advanced Materials*, **15**, 300–303 (2003).

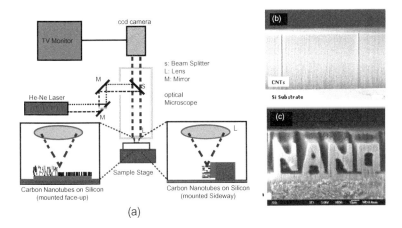

Figure 8.33. (a) Schematic of focused laser beam fabrication system. (b) Cross-Section SEM image of CNTs grown on Si substrate. (c) 3D NCT structures created using the focused laser beam technique.

The length of the CNTs removed depends on the power of laser beam used. In addition, as shown Fig. 8.33(a), the CNT samples can be mounted either face-up or sideways. In this way, we can fabricate unique three-dimensional (3D) structures made of CNTs. This process bears a close resemblance to pruning of hedges into unique structures. During the cutting of the CNTs, the same objective lens (L) is used to collect reflected light from the sample for viewing purposes. The reflected optical image is captured by a CCD camera that is coupled to a monitor. In this way, one can inspect the structures created.

The starting sample is typically a sample with aligned array of CNTs on a substrate. Figure 8.33(b) shows a cross-sectional SEM image of such a CNT sample. The multi-walled CNTs are uniform in length, each with a diameter of about 30 nm. When the CNTs are exposed to the focused laser beam, the CNTs disintegrate readily. During the experiment, the laser beam is kept stationary and the sample is moved by way of a computer-controlled stage with respect to the laser beam. In this way, a wide variety of microstructures can be created out of the aligned array of CNT forest. Starting from CNTs grown on the Si substrate, we can use the focused laser beam to cut out a wide variety of 2D and 3D structures as shown in Figs. 8.33(c) and 8.34.

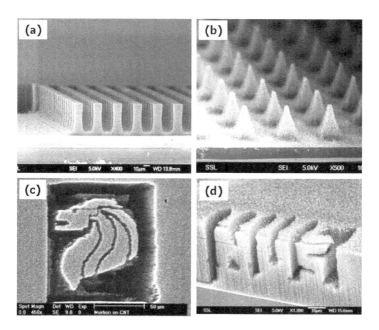

Figure 8.34. SEM images of some of the 3D structures created using the laser-pruning technique (from author's lab).

Exercises

8.1 Figures (a) and (b) are SEM images of two different systems taken using the same SEM. The scale bar in (a) is equal to 1 micron. It is known that the magnification of the image shown in (b) is twice the magnification shown in (a). What is the distance between two of the nanoparticles as shown in (b)? (Note that this distance is marked by the arrow shown in (b).)

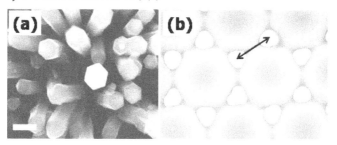

8.2 The following figures represent the chemical composition map of a sample (Pb = Lead, Ti = Titanium, B = Boron).

(i) Assuming that you use a Scanning Electron Microscope to take a picture of this sample using the backscattered electrons (BE) detector, sketch the resultant BE image. (ii) Suppose you insert an X-ray detector into the SEM system and set the detector to detect only X-ray photons with energy of 4.5 keV, sketch the resultant X-ray map. (Note that for this part, you are required to find out from other sources the energies of X-ray photon emitted by different elements.)

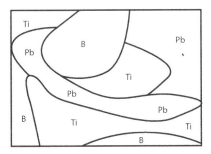

8.3 In STM, the tunneling current depends exponentially [$I = I_o \exp(-2ks)$] on the distance, s, between the tip atom and the surface atom (center-to-center distance). Let I_A be the tunneling current due to tunneling between the atom (A) and the surface atom. Let I_B be the tunneling current due to tunneling between the atom (B) (one level higher than tip atom) and the surface atom. Calculate the ratio I_B/I_A. (Assume $k = 5 \times 10^9$ m^{-1}, and diameter of the atoms is 0.5 nm and s = 2 nm.)

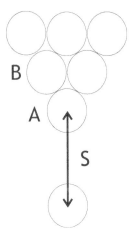

8.4 The picture on the left shows atoms G and atoms H on the surface of a sample. The energy band diagram on the right shows the energy level of a STM tip and the sample. Sketch the expected STM image in each case if we apply (i) a bias of $-4V$ to the tip and no bias to the sample; (ii) a bias of $-1V$ to the tip and a bias of $1V$ to the sample; (iii) a bias of $2V$ to the tip and a bias of $-2V$ to the sample; (iv) no bias to the tip and a bias of -5 V to the sample.

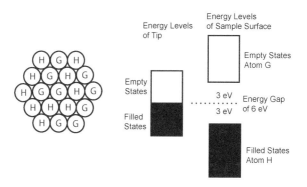

8.5 A microsphere (diameter 1 μm and marked **I**) was attached to an AFM cantilever for the measurement of interaction force between the sphere and another sphere (diameter 1 μm and marked **II**) on a substrate as shown in the following figure. Given that the two spheres interact with a force (unit N) of $F(r) = -2/r$ where r (unit μm) is the centre-to-centre separation between the spheres. Given that force constant of the cantilever is 1 N/μm and $\mathbf{H} = 5$ μm, what is the deflection of the cantilever. (Ignore the force contribution due to other spheres.)

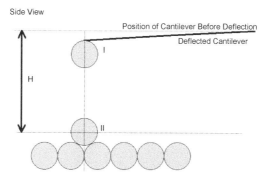

8.6 A colloidal system consists of aqueous suspension of two polystyrene spheres with diameter of 1.5 μm. Sphere **(1)** is optically trapped and sphere **(2)** is free. Water flow from left to right at a constant velocity of v. The following figure shows consecutive images of the colloidal system. Sphere **(1)** remained optically trapped throughout the experiment. (Note that scale bar is 10 μm and the viscous drag force acting on a sphere is given by F= $6\pi\eta$rv where η = 0.001002 Ns/m^2. (Ignore Brownian Motion.)

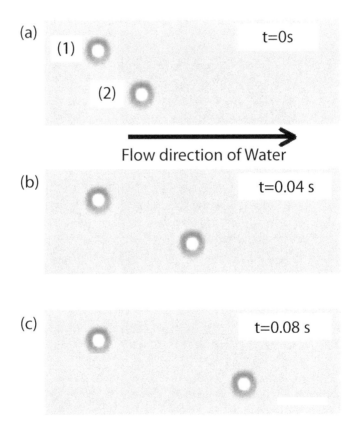

(a) (1) (2) t=0s

Flow direction of Water

(b) t=0.04 s

(c) t=0.08 s

(i) Calculate the viscous drag force acting on sphere (I).
(ii) If the optical tweezers exert a maximum force of 10 pN, what would be the terminal velocity?

Chapter Nine

Future Trends

So where will nanotechnology lead us in the coming years? Will nanotechnology become a mature and well-accepted technology, or will there be disappointment due to unrealised promises? Arie Rip, in his article *The Folk Theories of Nanotechnology*,[1] suggests that nanotechnology could follow a narrative of initial enthusiasm followed by subsequent disillusionment. The Gartner Group[2] depict this *cycle of hype* as initially accompanied by a flurry of publicity and unrealistic expectations, leading to a *peak of inflated expectations*. This is inevitably followed by disappointment and loss of public interest as the technology falls into a *trough of disillusionment*. Only then does the technology start to deliver, with a *slope of enlightenment* leading to a *plateau of productivity*, in which the technology does deliver real benefits, albeit less dramatic than those initially promised in the first stage of the cycle. Rip identifies the key issue as the degree to which it is regarded as acceptable to exaggerate claims about the impact of a technology. He observes a dichotomy in strategies between the USA and Europe, with advocates of nanotechology in Europe making much more modest claims, thus positioning themselves better for the aftermath of a bubble bursting.

Nevertheless, the concern of most nanoscientists is what real impact nanotechnology can deliver to improve the lives of ordinary people. How far are we away from that *plateau of productivity*? Since nanotechnology encompasses a diverse range of

---

[1] Arie Rip, "The Folk Theories of Nanotechnology", *Science as Culture* **15**, 349 (2006).
[2] http://www.gartner.com/

Science at the Nanoscale: An Introductory Textbook
by Chin Wee Shong, Sow Chorng Haur & Andrew T S Wee
Copyright © 2010 by Pan Stanford Publishing Pte Ltd
www.panstanford.com
978-981-4241-03-8

technologies from nanoelectronics to drug delivery, there are early indications that a few nanotechnology products are beginning to realize their potential. In this final chapter, we survey the impact of nanotechnology on society, particular on the developing world where it is needed most, and highlight some areas in which nanotechnology has had substantial success.

9.1 NANOTECHNOLOGY AND THE DEVELOPING WORLD

Researchers at the University of Toronto Joint Centre for Bioethics and the Canadian Program on Genomics and Global Health[3] (Toronto, Canada) show that several developing countries are already harnessing nanotechnology to address some of their most pressing needs. They identify and rank the ten applications of nanotechnology most likely to benefit developing countries, namely:

1. Energy storage, production, and conversion (Novel hydrogen storage systems, photovoltaic cells, organic light-emitting devices . . .)
2. Agricultural productivity enhancement (Nanoporous zeolites for slow-release of water and fertiliser, nanocapsules for herbicide delivery . . .)
3. Water treatment and remediation (Nanomembranes for water purification, desalination and detoxification, nanosensors for the detection of contaminants and pathogens . . .)
4. Disease diagnosis and screening (Nanolitre lab-on-a-chip, nanosensor arrays, quantum dots for disease diagnosis, magnetic nanoparticles as nanosensors . . .)
5. Drug delivery systems (Nanocapsules, liposomes, dendrimers, buckyballs, nanobiomagnets and attapulgite clays for slow and sustained drug release . . .)
6. Food processing and storage (Nanocomposites for plastic film coatings used in food packaging, antimicrobial nanoemulsions for decontamination of food equipment, packaging . . .)

---

[3] F. Salamanca-Buentello *et al.*, "Nanotechnology and the developing world", *PLoS Medicine*, **2**(5), 0300–0303 (2005).

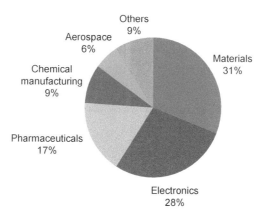

Figure 9.1. Estimates of the US$1.1 trillion nanotechnology market by 2010–2015 (Source: National Science Foundation, USA).

7. Air pollution and remediation (TiO_2 nanoparticle-based photocatalytic degradation of air pollutants in self-cleaning systems, nanocatalysts for catalytic converters, nanosensors for detection of toxic materials ...)

8. Construction (Nanomolecular structures for water-proof asphalt and concrete, heat-resistant nanomaterials to block ultraviolet and infrared radiation, self-cleaning surfaces ...)

9. Health monitoring (Nanotubes and nanoparticles for glucose, CO_2, and cholesterol sensors, and for in-situ monitoring of homeostasis ...)

10. Vector and pest detection and control (Nanosensors for pest detection, nanoparticles for new pesticides, insecticides ...)

It is important for the international community to work together to find ways to accelerate the use of these nanotechnologies by less industrialised countries to meet their critical sustainable development challenges. Figure 9.1 shows the NSF projected nanotechnology market by 2010–2015; it is clear that materials and electronics will be the first areas that nanotechnology will find commercialisation opportunities.

9.2 BEYOND MOORE'S LAW

There will be limits to Moore's Law and how far the current CMOS-based silicon technology can be pushed. These limits could arise from fundamental physics or materials science issues, or from engineering problems such as chip overheating. The basis for propagating Moore's law is the International Technology Roadmap for Semiconductors,[4] a document which maps out the research and development required to deliver the incremental improvements in CMOS technology. It is becoming increasingly clear that a demanding series of linked technological breakthroughs are needed as time progresses, and a roadblock may be reached sometime around 2020.

Richard Jones envisions three possible outcomes beyond 2020.[5] The first is that these problems will be solved, and Moore's law will continue through further incremental developments. The history of the semiconductor industry tells us that this possibility should not be taken lightly; the ingenuity of engineers and scientists to overcome seemingly insurmountable technical problems has kept Moore's law on track for forty years. The second possibility is that a fundamentally new technology, quite different from CMOS, will be developed, giving Moore's law a new lease of life. Currently, the likely contenders appear to be spintronics, quantum computing, molecular electronics, or graphene electronics. Although there has been a lot of excellent science being reported in these fields, none of these developments are close to commercialisation yet.

The third possibility that Jones proposes is that we enter into a period of relatively slow innovation in hardware, but this would not necessarily mean that there would be no developments in software. On the contrary, as raw computing power gets less abundant, human ingenuity in making the most of available power is likely to have a greater impact. The economics of the industry would change dramatically, and since the hardware development cycle would lengthen, the huge capital cost of wafer fabrication plants would be spread over a greater period of time, leading to the chip business becoming increasingly commoditised.

9.3 SPINTRONICS AND SURFACE CHEMISTRY

The 2007 Nobel prize for Physics was awarded to Albert Fert and Peter Grünberg "for their discovery of Giant Magnetoresistance

---

[4] http://www.itrs.net/
[5] Richard A. L. Jones, *Soft Machines: Nanotechnology and Life*, OUP (2004), and his blog: http://www.softmachines.org/wordpress/

(GMR)". GMR is a quantum mechanical effect observed in thin film structures composed of alternating ferromagnetic and nonmagnetic metal layers. The effect manifests itself as a significant decrease in electrical resistance in the presence of a magnetic field. The Nobel Foundation's press release interestingly concludes with this paragraph[6]:

> *"The GMR effect was discovered thanks to new techniques developed during the 1970s to produce very thin layers of different materials. If GMR is to work, structures consisting of layers that are only a few atoms thick have to be produced. For this reason GMR can also be considered one of the first real applications of the promising field of nanotechnology."*

The discovery was made in 1988, and it was realised that this effect would make it possible to manufacture very sensitive magnetic read heads for hard disks. In a remarkably short time, GMR technology was incorporated into their hard drive head and launched onto the market by IBM in 1997. This invention is responsible for the ultra-high density disk drives in MP3 players, digital video recorders, and computer notebooks (see Fig. 9.2). The IBM GMR website[7] proudly reads: *"The Giant Magnetoresistive*

Figure 9.2. Left: A hard disk drive; Right: Close-up of a disk drive head resting on the reflective disk platter together with its mirror image [Left image: Source – http://commons.wikimedia.org/w/index.php?title=Image:Hard_disk.jpg&oldid=11516558, Right image: Image courtesy of Mr. Andrew Magill. Copyright © 2006 by Andrew Magill www.ominoushum.com]

---

[6] http://nobelprize.org/nobel_prizes/physics/laureates/2007/press.html
[7] http://www.research.ibm.com/research/gmr.html

Figure 9.3. STM images of organic molecules self-assembled on surfaces, an important topic of study in surface and nanoscale science (from author's lab).

Head: A giant leap for IBM Research. To some people, 10 years = a decade. To IBM Research, 10 years = a revolution."

The 2007 Nobel prize for Chemistry was awarded to Gerhard Ertl "for his studies of chemical processes on solid surfaces". Using the powerful tools of nanoscale surface science (such as STM, see Fig. 9.3), he pioneered groundbreaking studies in surface chemistry, which help us to understand processes such as corrosion, catalysis and even semiconductor fabrication. Surface chemical reactions on catalytic surfaces play a vital role in many industrial operations, such as the production of artificial fertilizers, a vital ingredient in feeding the world's population.

The 2007 Nobel prizes in Physics and Chemistry highlight the importance and multidisciplinary nature of nanoscience and nanotechnology. Indeed, one could argue that these were both discoveries in surface and interface science, the nanoscale component in both cases being in one dimension, normal to the surface or interface. GMR was a discovery in fundamental solid state physics that was realized to be useful, and quickly commercialised. The chemistry prize, on the other hand, rewarded the achievements of surface science used to better understand processes that are already technologically important. This knowledge can in turn be used to improve these processes, or to develop new ones.

9.4 CARBON ELECTRONICS

The idea of using carbon or organics in electronics goes back several decades. We have mentioned that silicon CMOS technology may reach its miniaturisation limits in a few decades, and we may have to look for an alternative technology paradigm. One question sometimes asked is: Will carbon ever replace silicon in electronics? We summarize here the developments in organic electronics, molecular electronics, carbon nanotube electronics, and most recently, graphene electronics.

Organic electronics, or plastic electronics, is a branch of electronics that deals with conductive polymers, plastics, or organic molecules. The pioneers of highly-conducting organic polymers are Alan J. Heeger, Alan G. MacDiarmid, and Hideki Shirakawa, who were jointly awarded the Nobel Prize in Chemistry in 2000 for their 1977 discovery and development of oxidized, iodine-doped polyacetylene.

Conducting polymers are lighter, more flexible, and less expensive than inorganic conductors. Besides being a desirable alternative in many applications, they also open up the possibility of new applications that would be impossible using inorganics. Organic light-emitting diodes (OLEDs) have already been commercialised, and are being used in television screens, computer displays, portable screens, advertising, and signboards. An advantage of OLED displays over traditional liquid crystal displays (LCDs) is that OLEDs do not require a backlight to function. They draw far less power and, when powered from a battery, can operate longer. OLED-based display devices also can be more easily manufactured by printing methods, as compared to current LCD and plasma display manufacturing technologies.

New applications in organic electronics include smart windows and electronic paper. Smart window technology allows home owners to block either all or some light by simply turning a knob or pressing a button. This type of light control could potentially save billions of dollars on heating, cooling and lighting costs. Electronic paper (or *e-paper*) mimics the appearance of ordinary ink on paper. Unlike traditional displays, e-paper can be crumpled or bent like traditional paper. Imagine the amount of trees saved if our newspapers and books could be easily downloaded into our personal e-paper!

The ultimate goal in device miniaturisation is to make devices with a single molecule. Molecular electronics (or *moletronics*) is an interdisciplinary field that spans physics, chemistry, and materials science (cf. Section 1.3). The unifying theme is the use of molecular building blocks for the fabrication of electronic components.

The concept of molecular electronics has created much excitement among scientists and technologists due to the prospect of size reduction in electronics offered by the molecular-level control of properties. Although the original molecular rectifier was predicted as far back as 1974,[8] a commercialisable single molecule device has yet to be demonstrated.

The past few decades has seen the discovery of fascinating new allotropes of carbon. The *fullerenes*, discovered in 1985 by Robert Curl, Harold Kroto and Richard Smalley, are a family of carbon allotropes named after Richard Buckminster Fuller and are sometimes also called *buckyballs*. Kroto, Curl, and Smalley were awarded the 1996 Nobel Prize in Chemistry for their discovery of this class of compounds. Fullerenes are molecules composed entirely of carbon, in the form of a hollow sphere, ellipsoid, or tube. Cylindrical fullerenes are called *carbon nanotubes*. Fullerenes are similar in structure to *graphene*, which is a single 2D sheet of graphite made up of linked hexagonal rings (Fig. 9.4).

Graphene is the latest low dimensional material that has caught the attention of scientists due to its novel properties. Graphene has a linear energy–momentum dispersion relation similar to that of a photon in free space; hence its electrons behave as relativistic massless Dirac fermions. The low-energy electronic band structure of single layer graphene is unique, consisting of conduction and valence bands that meet at the charge neutrality level. As such, graphene has a zero band gap, whereas the semiconductors used in electronic devices typically have band gaps of between 1 and 2 eV. Challenges to graphene electronics therefore include

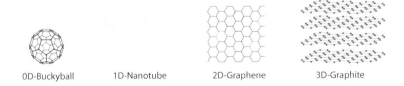

0D-Buckyball   1D-Nanotube   2D-Graphene   3D-Graphite

Figure 9.4. Graphene is a 2D building material for carbon materials of all other dimensionalities. It can be wrapped up into 0D buckyballs, rolled into 1D nanotubes or stacked into 3D graphite.

---

[8] A. Aviram and M. A. Ratner, *Chemical Physics Letters*, **29**, 277 (1974).

creating a bandgap and methods of doping graphene, as well as the large scale preparation of high quality graphene.

The story of nanoscience and nanotechnology still continues, but this book will have to conclude here. New discoveries in nanoscience await us in the years ahead, and a few of these may develop into nanotechnologies that could truly transform our lives in the future.

Further Reading

Soft Machines: Thoughts on the future of nanotechnology from Richard Jones: http://www.softmachines.org/wordpress/

D. Grundler, Spintronics, *Physics World*, April 2002, 39–43.

A. K. Geim and K. S. Novoselov, The rise of graphene, *Nature Materials*, **6**, 183–190, March 2007.

This page intentionally left blank

Index

Science at the Nanoscale: An Introductory Textbook
by Chin Wee Shong, Sow Chorng Haur & Andrew T S Wee
Copyright © 2010 by Pan Stanford Publishing Pte Ltd
www.panstanford.com
978-981-4241-03-8

T - #0999 - 101024 - C226 - 229/152/12 - PB - 9789814241038 - Gloss Lamination